T0293447

Culturally Tuning Change Management

Best Practices and Advances in Program Management Series

Series Editor
Ginger Levin

RECENTLY PUBLISHED TITLES

PgMP® Exam Test Preparation: Test Questions, Practice Tests, and Simulated Exams
Ginger Levin

Managing Complex Construction Projects: A Systems Approach
John K. Briesemeister

Managing Project Competence: The Lemon and the Loop
Rolf Medina

The Human Change Management Body of Knowledge (HCMBOK®), Third Edition
Vicente Goncalves and Carla Campos

Creating a Greater Whole: A Project Manager's Guide to Becoming a Leader
Susan G. Schwartz

Project Management beyond Waterfall and Agile
Mounir Ajam

Realizing Strategy through Projects: The Executive's Guide
Carl Marnewick

PMI-PBA® Exam Practice Test and Study Guide
Brian Williamson

Earned Benefit Program Management: Aligning, Realizing, and Sustaining Strategy
Crispin Piney

The Entrepreneurial Project Manager
Chris Cook

Leading and Motivating Global Teams: Integrating Offshore Centers and the Head Office
Vimal Kumar Khanna

Project and Program Turnaround
Thomas Pavelko

Culturally Tuning Change Management

Risto Gladden

CRC Press is an imprint of the
Taylor & Francis Group, an **informa** business
AN AUERBACH BOOK

PMP®, PgMP®, and *PMBOK® Guide* are registered trademarks of the Project Management Institute, Inc., Newtown Square, Pennsylvania, which is registered in the United States and other nations.

The Standard for Change Management® is a registered trademark of the Association of Change Management Professionals, Ovieda, Florida.

CRC Press
Taylor & Francis Group
6000 Broken Sound Parkway NW, Suite 300
Boca Raton, FL 33487-2742

Printed on acid-free paper

International Standard Book Number-13: 978-0-8153-8089-4 (Hardback)

Visit the Taylor & Francis Web site at
http://www.taylorandfrancis.com

and the CRC Press Web site at
http://www.crcpress.com

Dedication

To everyone who has ever ordered a cappuccino after 11 AM in Italy.

Contents

List of Figures

List of Tables

Preface

Context matters. I started thinking about contextual intelligence a few years ago, when a client bristled when I responded, "It depends," to a question he asked me about managing change. "It depends" was definitely not the answer he expected or wanted, and I mentally kicked myself as soon as I heard the words slip from my lips. No client wants to hear, "It depends." I knew that. But it was the truth. And as much as we'd like to tell ourselves otherwise, successful change management involves more than knowledge of methodologies, processes, and techniques. It requires understanding the context in which the change is taking place—and, of course, knowing what works in what situations. There is nothing wrong with the change management content or tools we have at our disposal, but the application of them does require a bit of critical thinking. It requires contextual intelligence.

If you've not heard the term before you might be wondering what I mean by *contextual intelligence.* Well, in a nutshell, contextual intelligence is a skillset that integrates the concepts of context, intelligence, and experience. Context consists of all the factors—internal, external, and interpersonal—that contribute to the uniqueness of a situation. Intelligence is our ability to transform data into information, information into knowledge, and knowledge into practice. Experience is measured by our ability to intuitively extract wisdom from different experiences. It is about knowing *how* to do something (technical knowledge) but being wise enough (based on intuition and experience) to know *what* to do. As a skillset, contextual intelligence is the ability to apply intelligence and experience to quickly and intuitively diagnose and interpret the context of a situation and then to use that new knowledge to exert influence—the implication being that contextually intelligent people can influence others regardless of the setting. And the concept of contextual intelligence may help explain what is missing when we flourish in one environment but not in another.

You probably know from experience that every change initiative is unique, and that taking that uniqueness into account is rudimentary to the success of your change management efforts. Most current change management methodologies do a fairly good job of emphasizing the importance of tailoring change management plans to align with the specific characteristics of the change (small or large, incremental or radical) and the organization (industry, know-how, technical competencies required, change competency, etc.). This is a really good start. But the context of a change is much more involved than knowing the specific work setting. It also includes other critical variables such as genders, geographies, and cultures (national, ethnic, and global), variables that most of the change management methodologies have skimmed over or neglected altogether. So, this brings me to another "intelligence"—*cultural intelligence.* Essentially, cultural intelligence is the ability to function and manage effectively in culturally diverse settings. It has to do with our capacity to adapt to unfamiliar cultural environments and then transform the experience gained during those cross-cultural encounters into knowledge and, ultimately, a global mindset. Some people like to call it "cultural mindfulness."

How do you feel about people, places, and things that are foreign to you? Most of us have grown up learning how to work with people who are like us. We have developed a uni-cultural lens that has helped us understand and interpret our surroundings. Wherever you come from, you will be accustomed to the views of how you were raised and your idea of "normal." But what happens when you find yourself working with people who are different from you, people who view the world through cultural lenses different from your own? Each culture has its own behavioral code, and the actions, gestures, and words we encounter in a cross-cultural setting can be easily misinterpreted, leading to misunderstanding and missed opportunities for cooperation. And things can get really tricky when you are tasked with implementing change in different cultural contexts. Studies show that cultures differ in terms of things such as decision making, problem solving, and change. So, it would be a good idea to match your change management strategy to the culture by diagnosing the contextual landscape through the appropriate cultural lens or lenses. To do this, you need to be able to, and have the desire to, adopt the perspectives, preferences, values, and needs of those who are culturally different from you. And this requires a good dose of cultural intelligence. Sound scary? The good news is that anyone who is motivated enough can cultivate cultural intelligence—and, if you're managing change across national and cultural boundaries, you should! Cultural intelligence is a skill that you can't do without in today's global economy.

Now, you may be thinking that the emerging global culture—driven by the information revolution, the greater ease of communication, and globalization—has made "national culture" irrelevant as a contextual variable. Well, it is certainly true that global culture has an effect on national culture. But cultures differ greatly in how much they have been affected by globalization. In fact, differences in openness to globalization may be down to cultural values at the national level. And while globalization may ultimately carry us closer to universal standards for work, we're not quiet there yet. There is still a very clear and present need to incorporate cultural aspects into everyday business and to develop strong, effective, and culturally intelligent change leaders.

Working globally has exposed me to different languages, value systems, and institutional environments. It has been one of the most broadening aspects of my personal and professional life and has instilled within me new ways of learning and responding to social–cultural differences. But working globally is no longer just about "where" we do business but "how" we do business. So, even when you're working domestically, chances are you're going to find yourself faced with cross-cultural interactions at some point. We all need to start thinking globally. When we think globally we are able to develop better communications, relationships, and understanding among colleagues, customers, and world partners. (See Sidebar "Never Order a Cappuccino After 11AM in Italy.")

Never Order a Cappuccino After 11 AM in Italy

Do you know the famous "cappuccino rule"? I didn't, until I made the mistake of ordering a cappuccino after lunch. I had just started working in the London office of a company headquartered in Venice, Italy. On my first trip to Venice, one of my Italian colleagues took me out for a lovely lunch.

After lunch, he asked, "Will you take a coffee?"

"Yes," I replied.

"What would you like?," he asked.

"Well, since I'm in Italy, I'll have a cappuccino," I replied enthusiastically.

My colleague's countenance clouded and his brow furrowed. I knew instantly that I had made some sort of a faux pas, but I didn't know exactly what.

"Oh. Should I not order a cappuccino?" I asked apologetically.

"No, no—if you want a cappuccino, you can have a cappuccino," he said still frowning. "But it makes Italians sick," he proclaimed with a look as though there were a bad smell under his nose.

In an attempt to recover from my offence, I asked him what an Italian would order. I learned that after lunch an Italian might order café espresso, lungo, or macchiato, but never, ever a cappuccino.

I learned to adapt my behavior and would now never order a cappuccino after 11 AM in Italy. Although, I must confess that I do take some perverse pleasure when I see other visitors to Italy make the same faux pas.

Aim of This Book

Plenty of change management approaches have gained wide acceptance, but they largely reflect Western-oriented, uni-cultural assumptions about what constitutes value and how to motivate people. Trying to apply them uniformly across geographies can be an exercise in frustration, if not a waste of time. There is certainly peril in this approach. Managing change across cultures requires contextual and cultural intelligence. We need to understand so many things better than we do. What do "They" perceive as change, and how do "They" respond to change? What forms of communication will "They" respond to? What is "Their" concept of self? The answers to these and other questions will vary from London to Moscow and from Moscow to Beijing. You should not assume that your technical change management knowledge will trump local customs and conditions.

Most of the change management approaches emphasize organizational culture as an important contextual variable, but neglect national culture as an equally important variable. So, I have written this book—drawing on personal experience as well as the vast and varied literature—with the hope that it will contribute to addressing this gap by giving you some food for thought and encouragement to be reflective and culturally mindful when you are planning and implementing change in culturally diverse contexts. I also hope that it will motivate you to move out of your own comfort bubble and seek out opportunities to become more culturally intelligent. Pushing your limits and challenging yourself is healthy, plus it can be a great boost to your self-confidence.

While this book is not an in-depth tool kit to managing change, I hope it will give you a meaningful overview of cultural differences and act as a guide to interpreting and adapting to different cultural behaviors. And for those of you tasked with leading change across borders, I hope this book will give you a better understanding of how culture influences the way people perceive and react to change, some guidance on how to develop your cultural intelligence, and some ideas about how you can plan and implement your change efforts in a more culturally mindful way. Maybe it will act as a catalyst for you to develop new skills, or find out that you had those skills in the first place but just haven't had the opportunity to use them before.

How the Book Is Organized

This book is organized into two parts. In the first part I cover the concepts of change, culture, and perspectives in order to establish context for the rest of the book. You might already be familiar with some of the concepts covered in these chapters, but you might define them differently to the way I am defining them here. For example, if you are a project manager, when you see "change management" you might immediately think "change control," which is not the definition of change management I am using.

In the second part of the book I examine the new competencies you need to successfully manage change in culturally diverse settings, as well as some suggestions for how you can develop these competencies. In this section, I also inspect some of the cultural dimensions that influence perceptions of and reactions to change in different cultural contexts and how this information can be used to develop culturally mindful change management plans.

— Risto Gladden
Englefield Green, Surrey, UK
9th July 2018

Acknowledgments

Nobody has been more important to me in the pursuit of this book than the members of my family—Tim, David, Susan, and Barney (my "chief happiness officer")—who put up with me throughout a trying process. I owe you all a heartfelt thank you for your daily support.

There are plenty of other people who helped bring this book to fruition, and I am grateful to all of them. Thank you to Dr. John Wyzalek at Taylor & Francis for seeing promise in this project, and to Theron Shreve and Susan Culligan at DerryField Publishing Services for patiently coaching me through the production process. I would especially like to thank Dr. Ginger Levin, who encouraged me to write this book in the first place and who has taught me more than I could ever give her credit for here.

Thank you also to the people who have inspired me. To my friends and colleagues around the world who sometimes push me out of my cultural comfort zone but who always help me to see things from a different perspective, and to Geert Hofstede and all of the other cross-cultural researchers whose work provided the foundation for the concepts in this book.

Above all, thank you Markus for just being you.

About the Author

Risto Gladden, PMP®, PgMP®, MBCS CITP, PCAI

Risto Gladden is a UK-based change management consultant and Prosci Certified Advanced Instructor (PCAI). He has lived and worked in several countries and has studied a number of languages, which has contributed to his interest in culturally tuning change management.

Over the past 25 years, Risto has been involved in shaping and delivering programmes of change for organizations in the UK, US, Europe, Russia, South Africa, and Latin America. His practical experience is underpinned by degrees in International Relations, Spanish, Psychology, and Project Management, and certifications from globally recognized professional organizations including Prosci, Association of Change Management Professionals (ACMP), Project Management Institute (PMI), British Computer Society (BCS), and Scrum Alliance, among others.

Risto was a member of the ACMP Standards Working Group tasked with the development of first global standard for change management, and he served as chair of the Credential Governance Committee.

Part I

Change, Culture, and Perspective

*We don't see things as they are, we see them as we are.**

— Anonymous

What is change? What is culture? The way you answer these questions depends on your perspective. If you're a project management professional, you probably think of change control—controlling the scope, time, cost, and quality of a project—when you hear the word "change." But if you're a change management professional, you probably think of "people" when you hear the word "change." And regardless of whether you are a project management or a change management professional, you probably think of "organizational culture" first when you hear the word "culture." What we believe or how we feel about change and culture is strongly influenced by our perspective and the way we take in and process information. So, before we can look at managing change across cultures, we need to have a common understanding of "change" and "culture."

We all have biases, manufactured by our own experiences. They can affect our memory, motivation, decision making, and how we perceive events and evaluate groups. On the whole, biases can be helpful when they enable us to make quick judgments and decisions, but they can also be bad for business

* https://quoteinvestigator.com/2014/03/09/as-we-are/

when they hinder us from considering valuable information and options unintentionally, leading to poor choices, bad judgments, and erroneous insights. In the workplace, cultural bias can unintentionally blind us to why our co-workers, employees, or clients from cultural backgrounds different from our own respond to change the way they do. Our assumptions about what employees are like and how they are socialized to behave depend on our own perceptions of how the society is characterized in terms of these value dimensions.

Did you know that 74 percent of firms in the 2017 Fortune Global 500 were based outside of the United States? In fact, the number of US companies on the list dropped by two, and the number of Chinese companies increased by six. Between 1990 and 2012, the number of multinational companies increased from 3,000 to more than 100,00, with 900,00 affiliates. This suggests that the cross-border flow of goods, services, and know-how is increasing, and this has far-reaching implications for change management.

Successful change management involves more than knowledge of techniques and technical skills. It requires understanding the context in which we're operating—knowing what works in what situation. When we are able understand and let go of our fixed attitudes and approaches to change management, to look at things from a different perspective, we will recognize the value of and be more open to developing change management strategies with greater cultural sensitivity. Managing change across cultures requires us to understand comfort zones and to know that cultural differences exist without assigning values (better or worse, right or wrong) to those differences.

Chapter 1

Change Management Is Essential . . . and Sometimes Messy

Nothing stays the same in the business world,
and sometimes employees have a hard time with that.[*]

— Peter Economy

What do you think of when you hear the word "change"? Change means different things to different people, and the management literature itself contains a mind-boggling variety of understanding and approaches to organizational change. This is probably because the study of organizational change is interdisciplinary and draws from so many different fields (management, psychology, sociology, anthropology, political science, economics, etc.), and the agendas and preferences of different researchers naturally present contradictions. "Understanding change means coming to terms with the lessons from numerous philosophies and their messy interconnections" (Smith and Graetz, 2011). But, to establish a common context, let's simply define change management as a framework for affecting some sort of change within the organization and

* https://www.inc.com/peter-economy/resolved-you-will-help-yur-employees-deal-with-change-in-2014.html

for managing the effects of those changes on the organization and its human resources.

Change management isn't new, really. We've been managing change of one sort or another for millennia. But it wasn't until the 1960s that change management began to emerge as a recognized discipline, in the form of *organizational development.* Back then it was primarily concerned with interventions to support gradual modifications within an organizational framework (Bennis, 1969) or "incremental" change. Today, organizational change is more constant, complex, and dynamic than it used to be, and there is an increasing frequency of "transformational" change involving adjustments to the basic governing rules of the organizational framework and shifts in the attitudes, beliefs, and cultural values within the organization (Bartunek, 1988). And as transformational change has taken center stage, there has also been a growing recognition of the uniqueness and contextual richness of each change initiative and an increased emphasis on "soft" issues (motivation, leadership, culture, etc.) and sociological approaches based on "changing" rather than the change.

1.1 Don't Forget the People

Many change initiatives are doomed from the beginning. Why? Well, in a large way, it's because companies make the mistake of worrying mostly about the "change" rather than about "changing." They spend most of their time on the change (new product, new technology, new market, etc.) and very little time preparing the people who will have to change the way they work, act, or think in order for the change to be successful. Organizations are made up of people, and they are the ones who will either embrace or resist the change. Organizations don't change, people do—and this is a bit of a conundrum, because while people are crucial to the success of organizational change, they can be the biggest obstacle to achieving it. What happens if the only way an organization can achieve its goals is by persuading tens or hundreds or thousands of its people to change the way they work? It isn't enough to simply tell them they'll *have to* change the way they work. For change efforts to be successful, we need to get companies to boost their commitment to interventions aimed at ensuring employees are change ready.

Most organizational change involves some sort of adjustment to existing behaviors and ways of working or, in the case of transformational change, a reframing of attitudes, beliefs, and cultural values. It involves something old stopping, something new beginning. During the transition from the old to the new, a gradual psychological reorientation happens inside us as we try to adapt to the change. It's often this transition, rather than the change, that people resist. Some people are naturally more resilient during change, but for others the transition can be difficult, and their personal adjustment to the change can take longer. And let's face it—not all employees respond to change with a positive attitude. Reasonable-seeming people can sometimes be totally irrational, depending on the change. Given this state of affairs, the transitional period can be both a time-consuming process and risky—and things can get messy. But when change efforts fail, it's usually because we haven't paid enough attention to managing the transition. People will not change unless and until they are psychologically ready to.

There are many different types of change, and there are different approaches to managing it. Most organizational change is driven through projects, and for this reason, many people assume project management and change management are the same thing—they aren't. They are complimentary disciplines, but there is a fundamental difference between the project management and the change management perspective of "change management." From the project management perspective, "change management" is often understood to refer to "change control," which, if you are familiar with project management, you'll know is all about managing the project's baselines of scope, time, cost, and quality; how change requests will be created, approved, and managed throughout the life of the project; and so on.

In fact, the Project Management Institute's *PMBOK® Guide,* (PMI, 2013, p. 97) refers to the change management plan as the part of the project management plan that provides direction for managing the change control process and documents the formal change control board. From the change management perspective, "change management" is about adoption and sustainability of the change, whereas project management is about outputs (new product, new technology, etc.).

Change management is about converting project outputs into outcomes and benefits (improved performance, cost reduction, etc.). The emphasis is on transitioning individuals, teams, and organizations from the old way of doing things to the new way. Until recently, organizational change management has been fairly underrepresented in the project management literature (Hornstein, 2015). (See Sidebar "Change Management and Organization Project Management" on next page.)

Projects help organizations achieve strategic goals when the outputs provide business value, but more often than not they do not adequately address the impact on people and processes. Think about it. After a change (output) has been implemented, it is not uncommon for requests to come pouring in to undo some of what has just been delivered or for people to look for ways to return to the old methods of doing things. Take, for, example the organization that invested substantial time and resources in designing and implementing a workspace solution

Change Management and Organizational Project Management

Organizational Project Management (OPM) is an adaptive approach for fitting the capabilities of portfolio, program, and project management to the context and needs of the organization for the delivery of organizational strategy.

Change management integrates and aligns people, processes, structures, cultures, and strategy. In the context of OPM, change management helps organizations successfully drive their strategies through portfolio, program, and project management.

Best Industry Outcomes research (Crawford and Cooke-Davies, 2012) shows that fewer than 20% of organizations have change management capabilities. The research also indicates that organizations that have some change management capabilities do not necessarily use those capabilities and that there is a strong correlation between underutilization of change management and an increase in project failure.

Sources: PMI (2013); Crawford and Cooke-Davies (2012).

that involved hot-desking. The employees who were going to be most impacted by the change were not consulted about the solution before it was implemented, and most of them were infuriated when they learned they would no longer have their own desk. In a masterstroke of resistance, the employees found a way to revert to status quo. They simply had the administrative assistants add each desk in the building to the conference room reservation system so they could reserve their desks on a rolling 30-day basis. So much for hot-desking, and so much for the benefits the organization expected to get from more "casual collisions" and collaborative working. Why on earth do organizations like this invest so much time, money, and resources in projects if they aren't going to reinforce and sustain the change? Or worse—why do organizations end up investing more time and resources to turn around and undo some of the changes they have just delivered? Well, maybe because they are too focused on the shiny new change and not enough on the people who need to change the way they think and work in order for the change to be a success.

One of the key characteristics of any project is that it has a defined beginning and end. Although there may be some realization of benefits at the point of delivery of the project outputs, the expected business value is typically not achieved until those outputs have been fully adopted and embedded within the business. Sustainment typically involves ongoing activities that extend beyond the traditional scope of projects, and this is where change management comes into play. Change management is a structured approach for transitioning individuals, groups, and organizations from "where we are today" to "where we want to be in the future" to achieve intended business benefits. It helps organizations integrate and align people, processes, technologies, structures, and strategies. Project management and change management may seem to be separate in theory, but we need to integrate them in practice, because both are important to achieving organizational objectives (see Figure 1.1), and the success of one largely depends on the success of the other.

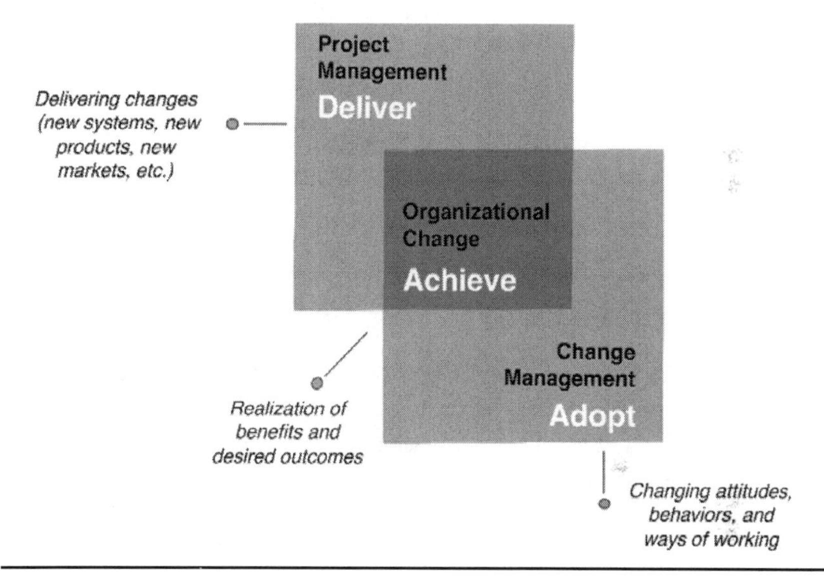

Figure 1.1 Complementary Disciplines of Project Management and Change Management

Effective organizational change encompasses both the "what" and the "how" of change. At a high level, we can divide organizational change into three interrelated parts:

- **Vision** (Leadership)
  Change leaders must create a clear and compelling vision of the desired future state and a clear expression of the reasons the change is needed and the goals to be achieved through the change.
- **Solution** (Project Management)
  The components of organizational project management (portfolio, program, and project management) play a significant role in translating the strategic vision of the organization into specific outputs. This includes

designing and delivering the solutions that fit into the organization's culture and enable the desired goals to be achieved.

- **Adoption and Sustainment** (Change Management)
  Change management is about understanding the people who will be impacted by the change and helping them make the transition from the current state to the future state. It includes preparing them to accept the outputs from the project, and, to a great extent, this is the bridge between projects and business as usual. The success of any change initiative lies in the benefit value for the organization. Benefits realization is dependent on sustainment, and sustainment typically involves ongoing activities that extend beyond the traditional scope of projects.

1.2 Leading and Managing Change

The notion that leading and managing change are different activities is not new. But what is the difference between leadership and management in the change process? We tend to think of change leaders as charismatic heroes, those executives at the top of the organization who envision, initiate, or sponsor strategic change. In contrast, we often think of change managers as facilitators and adapters, those middle managers who carry forward and build support for the change within the various business units. In other words, change leadership is about creating a vision of a change, while change management is about translating the vision into actions. The two roles are complementary, and in practice, they are sometimes indistinguishable, particularly in flatter organizations.

A Delphi-style panel of change agent experts recently identified and ranked in order of importance the sets of attributes they perceived to characterize the roles of change leaders and change managers (see Figure 1.2). In reality, some of the attributes required to lead and manage change are simply inseparable aspects of managerial work (Caldwell, 2003). For example, adaptability and flexibility, learning from others, and openness to new ideas are overarching attributes that are equally important for both change leaders and change managers. We could also add contextual intelligence—proficiently adapting knowledge and skills to different situations and environments—to the list of overlapping attributes.

Change is a way of life, and change-savvy leaders and managers have a key role to play in helping the people within their organizations embrace and adopt change. But successful change depends not only on how change agents lead and manage the change, but also on employee attitudes toward the change. Change agency needs to be a required skill for leaders, managers, and employees alike. Personal responsibility and accountability for change must be instilled at every level of the organization to strengthen the total organization's capability to change and build a culture in which change is accepted and expected.

Complementary Roles of Leading and Managing Change	
Key Attributes of Change Leaders	**Key Attributes of Change Managers**
1. Inspiring visions	1. Empowering others
2. Entrepreneurship	2. Team building
3. Integrity and honesty	**3. Learning from others**
4. Learning from others	**4. Adaptability and flexibility**
5. Openness to new ideas	**5. Openness to new ideas**
6. Risk taking	6. Managing resistance
7. Adaptability and flexibility	7. Conflict resolution
8. Creativity	8. Networking
9. Experimentation	9. Knowledge of the business
10. Using power	10. Problem solving

Source: Caldwell (2003)

Figure 1.2 Complementary Roles of Leading and Managing Change

1.3 Impact of Globalization on Change Management

Globalization is the most significant change taking place in today's work environment. Countries have become more interdependent through cross-national flows of goods and services, capital, know-how, and people. The ability to function effectively in different cultural contexts has never been more important, and employees who can work with peers from different cultures are becoming increasingly valuable in assisting companies integrate into the global market (Zoogah and Abbey, 2010; Lenartowicz and Johnson, 2007). But global teams are often confronted with the task of balancing Western project management and change management practices with local business customs in order to deliver results—often without an awareness of the impact of cultural factors on the outcome of the change.

There is a need to understand how national culture influences work and outcomes on multinational projects, and we'll explore this in the following chapters, along with some competencies and strategies for enabling change in a culturally mindful way.

Leading and managing change is highly culturally sensitive, and it is important to consider all of the contextual aspects of the change in order to identify any potential cultural trip wires. It is also important to allow for and promote the use of different approaches in different cultural contexts. In the following chapters, we'll take a look at the impact of culture on change and ways to approach change management in a more culturally mindful way.

Key Points

- Project management and change management are complimentary disciplines, and the success of one largely depends on the success of the other.
- Most organizational change requires some adjustment to existing behaviors and ways of working.
- For change efforts to be successful, organizations need to boost their commitment to interventions aimed at ensuring employees are change ready.
- Leading and managing change are different but overlapping change-enabling activities.
- Global teams need to understand how national culture influences work and change outcomes.

Want to Know More?

There is no shortage of sources of information on change management. *Making Sense of Change Management,* by Esther Cameron and Mike Green, is a good primer on the various theories, models, tools, and techniques of change management.

If you want to learn more about the relationship project management and change management, take a look at the Project Management Institute's *Managing Change in Organizations: A Practice Guide* (PMI, 2013).

The Standard for Change Management® was released by the Association of Change Management Professionals in 2014, and it is available for download from their website, https://www.acmpglobal.org/page/the_standard? (ACMP, 2014).

A Change Management Body of Knowledge (*CMBoK*) was published in 2013 and is available for purchase from the Change Management Institute (CMI), https://www.change-management-institute.com/buycmbok (CMI, 2013).

Prosci has been conducting research on best practices in change management since 1998 and represents the largest body of knowledge on the subject. The latest benchmark study is available for purchase from Prosci at www.prosci.com (Prosci, n.d.).

Chapter 2

Title Culture and Cultural Dimensions

*Determining national characteristics is treading a minefield of inaccurate assessment and surprising exception. There is, however, such a thing as a national norm.**

— Richard Lewis

Culture is an ubiquitous concept. It can be difficult to define. In fact, more than 160 different definitions of culture can be found in the literature (Kroeber, 1985). Most of the definitions seem to converge on the notion that culture is learned, it is associated with groups of people, and it encompasses a range of attributes including norms, values, shared meanings, and ways of behaving. More simply put, culture is a shared meaning system that influences how we think, act, organize, relate, and perceive. From an anthropological perspective, culture usually refers to societies defined in national or ethnic terms, but it can also be used to describe the concepts of organizational and group culture.

Globalization has accelerated over the past few decades, and so has research into international business. One area that has caught the attention of researchers is the effect of national culture on business processes and outcomes. But how much does culture really matter from a change management perspective? It matters a lot. Culture influences human behavior, and it plays an important role in how we interpret and respond to the work environment and change. It

* Lewis, 2006

subconsciously guides our behavior and our thoughts, and it often influences our sense of belonging, motivation, and effectiveness at work. Culture influences our expectations about things such as employment terms, working conditions, psychological contracts, and managerial roles. It can influence whether we expect management to lead in a more participative or authoritative way, the relative importance we place on employer reputation, and how we communicate and respond to organizational communication. Cultural differences are very important contextual variables, and our failure to account for them when implementing change can lead to embarrassing blunders and drag down business performance.

2.1 Organizations Are Not Cultural Islands

Since culture essentially refers to the way of doing things, it means that it also affects how a business is managed. An analysis of how the organization works can give us some meaningful insight into how the culture of the organization influences attitudes, behaviors, and ways of working within the workplace. It is not surprising then that most of the prevailing change management theories and models zero in on organizational culture as an important contextual variable that we need to consider when preparing for and implementing change. Consequently, it can be tempting for us to conceptualize the organization as a cultural island. But culture is a dynamic, multi-layered structure, and organizational culture is only one of several cultural spheres that influence what happens inside organizations. Organizational culture is not independent of the forces external to it. In fact, some of the variations we see within organizations can often be explained by similar variations at the societal or national level (Alvesson, 1993). It is probably more useful then to think of organizational culture as a point at which broader dimensions of culture converge (Meyerson and Martin, 1987).

Erez and Gati (2004) proposed a multi-level model of culture characterized by a hierarchy of levels nested within one another, where the innermost level is the cultural representation at the individual level nested within groups, organizations, nations, and the global culture (Figure 2.1). One cultural level affects change in other levels of culture in a top-down–bottom-up way.

This multi-level model of culture also depicts how the emerging global culture both affects the nested levels of culture below and is mutually affected by them. Through the top-down processes, we internalize the shared-meaning system of the society to which we belong, and its values are represented in our concept of self and identity. For example, collectivistic values are represented in an *interdependent* self, whereas individualistic values are represented in an *independent* self (Earley, 1994; Markus and Kitayama, 1991). Similarly, through the bottom-up processes of shared values, higher-level systems of culture are formed at the group, organizational, and national levels. Organizational culture itself is nested within the level of national culture. Of course, culture is not monolithic, and in some markets we may need to go beyond national culture in order to understand cultural diversities in terms of regional, ethnic, and even professional groupings within and across borders. It gets complicated.

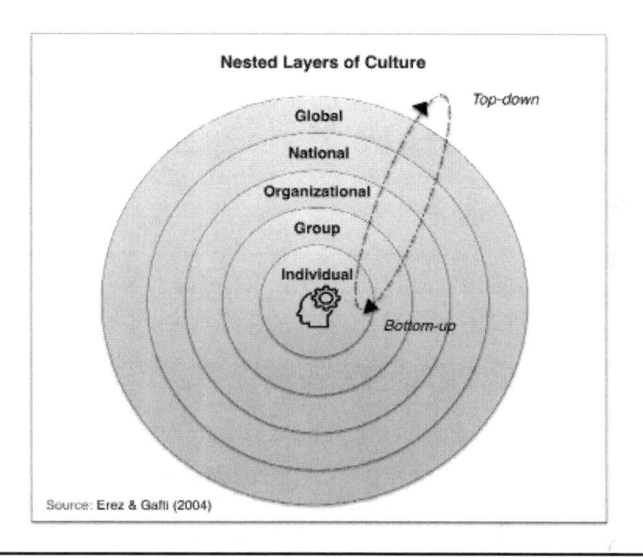

Figure 2.1 Multi-Level Model of Culture (Reprinted with permission from "A Dynamic, Multi-Level Model of Culture: From the Micro Level of the Individual to the Macro Level of a Global Culture," by Erez, M., and Gati, E. *Applied Psychology,* 53(2): 2004, p. 588. © John Wiley & Sons)

Edgar Schein (1992), a former professor at the MIT Sloan School of Management, proposed another dimension of organizational culture that reflects levels of visibility, ranging from the most visible to the least visible elements of culture (Figure 2.2).

Similarly, national culture consists of an external and visible level of behaviors and practices of the physical and social environment, a deeper middle level of expressed values, and an internal and invisible level of taken-for-granted basic assumptions that give direction and meaning regarding reality and human relationships. Certain aspects of cultural differences are easily observable, such as language, rituals, food, and music. Observable aspects of cultural differences abound at the external level, but values, beliefs, and basic assumptions may be less obvious and more difficult to interpret.

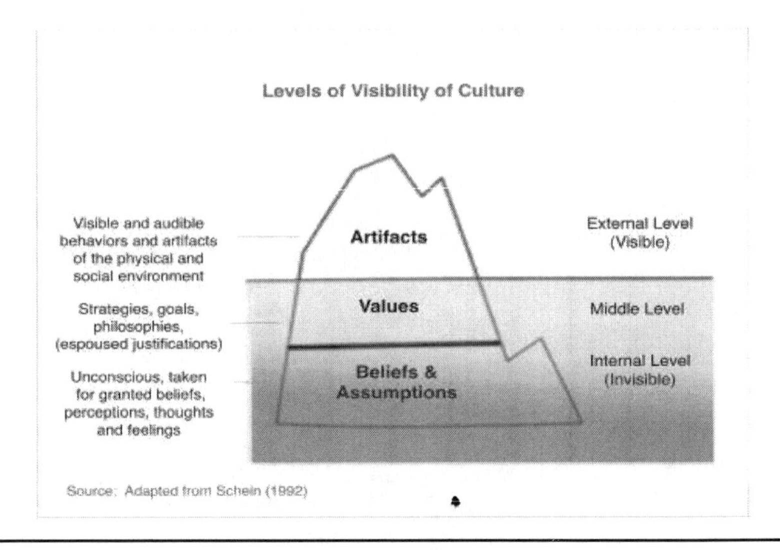

Figure 2.2 Levels of Visibility of Culture (Reprinted with permission from *Organizational Culture and Leadership,* by Schein, E. H. 1992, p. 28. © John Wiley & Sons)

We rarely think consciously about our own cultural identity, but national culture is the kind of culture that most strongly shapes our thinking, behavior, and perceptions of and reactions to the work environment and change. National culture characteristics are embodied in organizations, and in global settings, the nationality of the company often influences the culture of the organization. (IBM and McDonald's have a noticeably American feel to them wherever they operate in the world, Royal Dutch Shell has a distinctly Dutch ethos, and so on.) Things such as ethnicity, gender, age group, and profession are also all part of an individual's cultural identity and have profound effects on organizations. The challenge for change leaders is to integrate the expectations and work styles of multiple generations, nationalities, and subcultures to achieve a shared vision and shared strategies for the organization when it comes to change.

2.2 Impact of Globalization on National Culture

Culture is dynamic, and it can be adaptive in response to changes in society. Exposure to the global work environment can shape a global identity in which we develop a sense of belonging to a worldwide culture by adopting practices, styles, and information that are part of the global culture while continuing to hold onto our local identity, thus holding a bicultural identity (Arnett, 2002).

Foreign and domestic employees working in multinational and foreign-owned enterprises are usually socialized into the macro level of global culture, and they sometimes adopt a global mindset that enables them to adapt to their global work environment and to behave according to its core values. But how much are our identities changing as we become increasingly connected and mobile? Is a global identity more important than a national identity?

A 2016 Pollsters GlobeScan survey of 20,000 people in 19 countries found that more than half (56%) of those asked in emerging economies, such as Nigeria, China, Peru, and India, saw themselves first and foremost as global citizens rather than national citizens. But in Russia, only 24 percent of respondents saw themselves more as global citizens than citizens of their own country.

In some industrialized nations the concept of global citizenship appears to be heading in the opposite direction. In Germany, for example, only 30 percent of respondents saw themselves as global citizens. The global average among survey respondents was 51 percent. Majorities or strong pluralities in 16 countries described being a national citizen as the most important feature of their identity. And we can see a similar phenomenon within the European Union (EU), where some citizens of its Member States see themselves as Europeans first, while others identify more strongly or only with their national identity (see Sidebar "How 'European' do EU Citizens Feel?" on next page).

Global culture can be difficult to define, but research suggests that it appears to be dominated by certain Western values: freedom of choice, free markets, individualism, innovation and change, tolerance to diversity, and interdependence/connectedness (e.g., Erez and Gati, 2004). Cultures differ greatly in how much they have been affected by globalization, and differences in openness to it can often be explained by cultural values at the national level. A good fit between the global culture and a person's local identity enables them to effectively adapt to both.

Globalization and the emerging global culture may lead to some cultural convergence at the national level, but research suggests national cultural differences have remained fairly stable over time, despite globalization (Inglehart and Baker, 2000; Schwartz, 2008). "Many differences between national cultures at the end of the 20th century were already recognizable in the years 1900, 1800 and 1700, if not earlier. There is no reason why they should not play a role until 2100 or beyond" (Hofstede, 2011). A recent World Values Survey of almost 100 countries reflecting 90 percent of the world's population highlights the endurance of national cultural values even after factoring in globalization (see Sidebar "World Values Survey"). The world may be getting flatter, but it still has a lot of air in it when it comes to national culture versus global culture.

One of the great paradoxes of globalization is that, while it has made the world seem more connected in many ways, it has also increased the need for local contacts who understand the nuances of doing business in the local culture. The cultural heritage of most nations has a significant impact on its societal, economic, and political practices, which in turn influence the development of business practices. Our success in leading and managing change globally lies in how well we understand and can adapt to cultural differences while still preserving our core individual values and skills. An examination of national culture can give us a clearer understanding of a company's way of doing business, so let's take a look at some of the cultural dimensions that influence business practices and employee attitudes, behaviors, and ways of working.

How European Do EU Citizens Feel?

According to a long-running survey monitoring the continent's views on integration, 68 percent of people living in EU countries personally feel they are "European" (members of the European Union). The continent feels more positive about European identity.

In its Spring 2015 report, Eurobarometer reported that two-thirds (67%) of people living in the EU see themselves as "European."

Most people in the EU see themselves as "European." This has been the majority view since 1992, with the proportion ranging from 51 percent to 63 percent during that time. There are different shades of feeling European: Some people feel "national and European," others "European and national," and a small minority feel "European only."

But overall, the majority have felt European in some way every time this question has been asked. However, a substantial minority of Europeans feel "national only," in proportions varying from 33 percent to 46 percent over the last 20 years or so.

Eurobarometer has been surveying the views of Europeans since 1973 and gives a unique insight into how opinions and attitudes have changed over time. Surveys are carried out in all Member States of the European Union.

The feeling of being European varies significantly between Member States: 88 percent of citizens feel European in some way in Luxembourg, Malta (84%), Finland (81%), and Germany (81%). Fewer do so in the UK (58%), Italy (56%), Bulgaria (50%), Cyprus (50%), and Greece (50%).

This proportion falls to around a third in the UK (33%) and Ireland (34%), and to less than half in Greece (44%), Romania (46%), Portugal, and Bulgaria (both 48%).

Source: Standard Eurobarometer 83 Spring 2015. Retrieved from http://ec.europa.eu/public_opinion/index_en.htm

World Values Survey

The World Values Survey (WVS) is a global network of social scientists studying changing values and their impact on social and political life. Started in 1981, it consists of nationally representative surveys conducted in almost 100 countries, which contain almost 90 percent of the world's population. It is the largest non-commercial, cross-national investigation of human beliefs and values ever executed.

The Inglehart-Welzel Cultural Map, WVS-6 (2015)

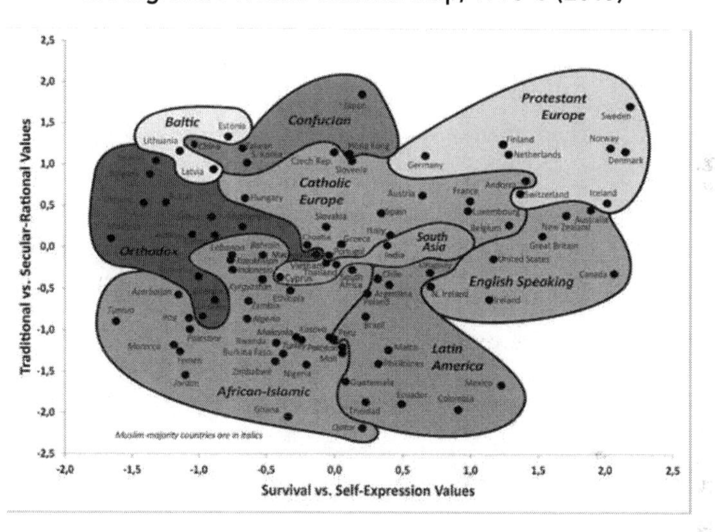

Traditional Values. Emphasis on religion, parent–child ties, deference to authority, and traditional family values.

Secular-Rational Values. Less emphasis on religion, traditional family values, and authority. Divorce, abortion, euthanasia, and suicide are seen as relatively acceptable.

Survival Values. Emphasis on economic and physical security. Relatively ethnocentric outlook and low levels of trust and tolerance.

Self-Expression Values. High priority to environmental protection, growing tolerance of foreigners, gays and lesbians, and gender equality, and rising demands for participation in decision making.

Sources: Reprinted with permission from Inglehart, R., C. Haerpfer, A. Moreno, C. Welzel, K. Kizilova, J. Diez-Medrano, M. Lagos, P. Norris, E. Ponarin and B. Puranen et al. (eds.). 2014. World Values Survey: Round Six–Country-Pooled Datafile Version: http://www.worldvaluessurvey.org/WVSDocumentationWV6.jsp. Madrid: JD Systems Inst.

2.3 Cultural Differences

The business environment is marked by great differences in national culture. These differences have been the focus of many studies attempting to understand the impact of national culture on various aspects of society, businesses, behaviors of individuals, etc. Drawing on Schein's model (Figure 2.2), theories of culture differ in their focus on the various "layers' of culture: visible and external (behaviors and practices), mid-level (values), and invisible and internal (basic assumptions). Most of the research on culture has focused on the middle level on the continuum between visible and invisible elements of culture.

Broadly, there are two basic approaches to the study of culture and social behavior: *etic* (culture-general) and *emic* (culture-specific) (Triandis, 1994). The etic approach tends to be quantitative and is concerned with identifying universal dimensions that underlie cultural differences. The emic approach is more qualitative and holds that "attitudes and behaviors are expressed in a unique way in each culture" (Chan and Rossiter, 2003).

A number of cross-cultural models have been developed, and two of them are compared in Table 2.1. These frameworks can be a good starting point for assessing cultural differences that may enable you to develop change management interventions that are more culturally suitable.

Table 2.1 Comparison of Hofstede Model and GLOBE Model

	Hofstede Model	GLOBE Model
Project Design	Dutch based	US based
Level of Analysis	National	National, societal
Basis	Empirical	Extension of Hofstede
Organizations Surveyed	1 (IBM and its subsidiaries)	951
Industries	Technology	Food processing, financial services, and telecommunications
Societies Surveyed	62	72
Respondents	Managers and non-managers	Managers
Cultural Dimensions	Power distance Uncertainty avoidance Individualism/ collectivism Masculinity/femininity Short-term orientation Indulgence	Power distance Uncertainty avoidance Humane orientation Collectivism I (organizational) Collectivism II (individual) Assertiveness Gender egalitarianism Future orientation Performance orientation

Definitions of culture vary according to the focus of interest, the unit of analysis, and the disciplinary approach (anthropology, psychology, sociology, etc.). This is significant in that studies of cultural differences adopt a specific definition and set of measurable criteria, which are always debatable. Of course, any research into culture and its impact in business and management studies is highly contentious. But there is a strong consensus that the key elements of culture include language, religion, values, attitudes, customs, and norms of a group or society. Figure 2.3 shows how the world's population is divided according to geography, language, and religion.

Let's take a closer look at some the theories and frameworks used to measure and compare national culture attributes.

2.4 Hofstede Model

Dutch social psychologist and management scholar Geert Hofstede is a chief architect of etic cross-cultural studies concerning the impact of national culture on business management. Between 1967 and 1973, he surveyed over 116,000 IBM employees in more than 70 national subsidiaries around the world. His analysis of their responses to questions about their work and job settings revealed cultural differences across four dimensions: power distance, individualism/collectivism, uncertainty avoidance, and masculinity/femininity (Hofstede, 1980). On the assumption that IBM's corporate culture was likely to be similar wherever it operated, Hofstede concluded that employees' differences in workplace values were likely to be a reflection of national culture differences.

In a subsequent study (Hofstede and Bond, 1991), Hofstede used a Chinese equivalent of his original survey developed by Chinese social scientists. Researchers identified an additional dimension representing Chinese values related to Confucianism. The study's co-author, Michael Bond, labeled the dimension Confucian Work Dynamism, but it was later re-labeled long-term/short-term orientation and was added as a fifth dimension to Hofstede's model. Then, in 2010, a sixth dimension—indulgence versus restraint—was added to the model.

This dimension essentially measures national levels of happiness and draws on the work of Bulgarian linguist Michael Minkov (2011) and the extensive World Values Survey (see page 17). Today, the Hofstede model of national culture consists of six dimensions (Figure 2.4) based on extensive research done by Hofstede, his son Gert Jan Hofstede, Minkov, and their research teams (Hofstede, Hofstede, and Minkov, 2010). The data they have gathered has enabled them to compute average scores on a scale from 0 to 100 along these six dimensions for each national culture involved in the study.

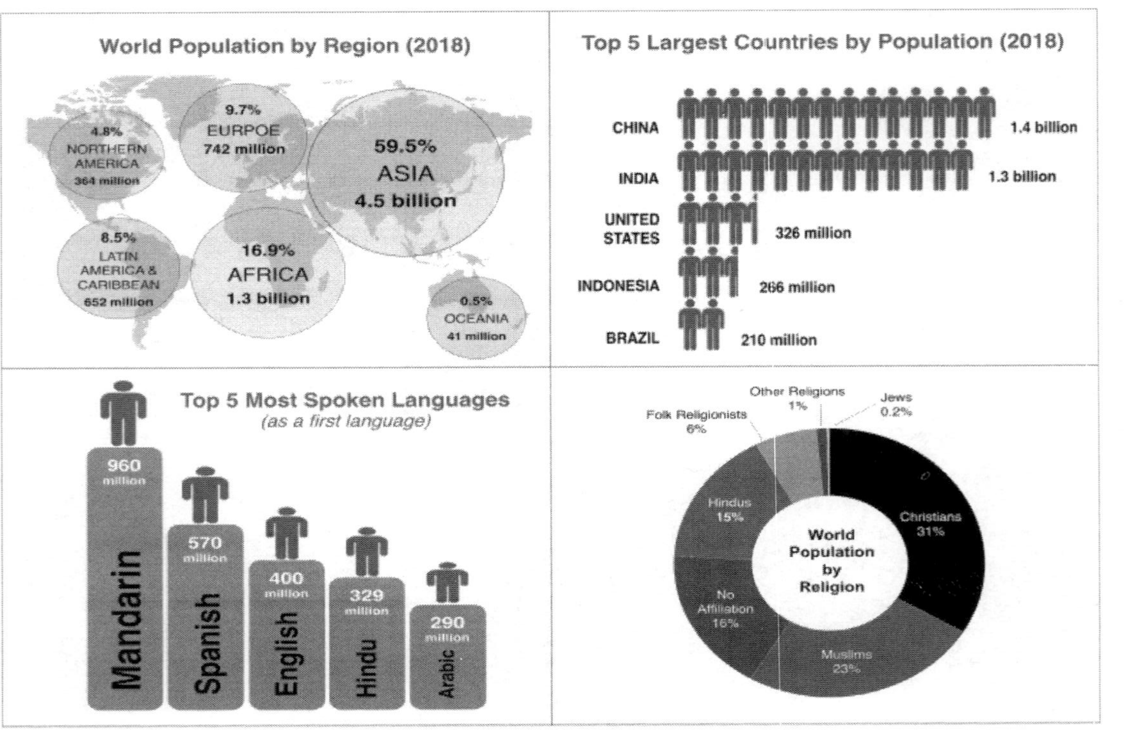

Sources: Ethnologue (2017 20th Edition); www.wolrdometers.info

Figure 2.3 World Population by Geography, Language, and Religion (Based on data drawn from Ethnologue)

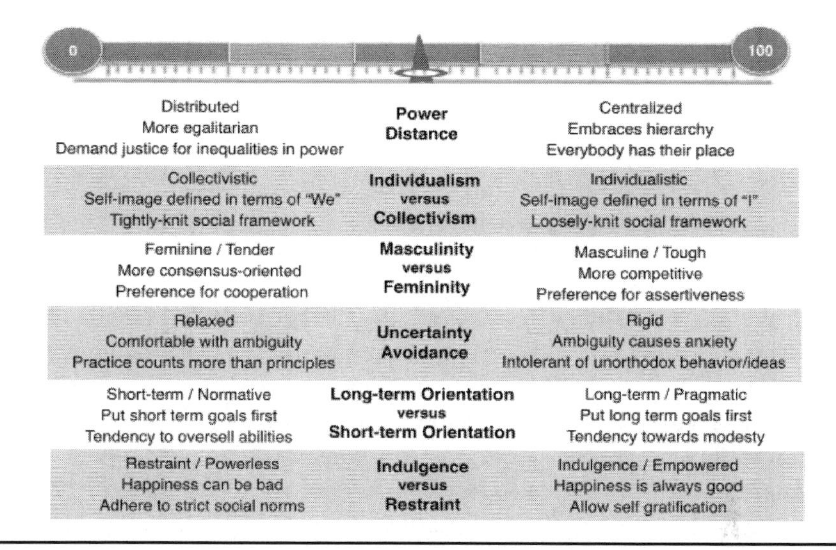

Figure 2.4 Hofstede's 6-Dimensions Model of National Culture (Reprinted from Geert Hofstede, Gert Jan Hofstede, Michael Minkov, *Cultures and Organizations, Software of the Mind,* Third Revised Edition, McGraw Hill 2010, ISBN 0-07-166418-1. © Geert Hofstede B.V. quoted with permission)

2.4.1 Power Distance

Would you confront your boss? If you're from Austria, you might. If you're from Malaysia, probably not. *Power distance* concerns the degree to which a culture accepts (and reinforces) the fact that power is distributed unevenly in a society. Put simply, people in some cultures accept a higher degree of unequally distributed power than do people in other cultures. High power distance cultures tend to accept power differences and show respect to their superiors. Lower power distance countries tend to be less comfortable with organizational rank and are often characterized by higher levels of participative decision making. For example, in low power distance cultures, subordinates might expect to be consulted, whereas in high power distance cultures, subordinates are more likely to expect to be told what to do. If you are from a low power distance culture and have to deal with someone in a high power distance culture, nothing much is going to happen without their boss's say so. From a change management perspective, people in high power distance cultures might appear more resistant to change because their dependence on superiors makes them less experienced in taking personal initiative to adapt to changes, so it is important to make sure you are talking to the right person or recognize that things may take a lot longer than you originally anticipated.

- **Countries with higher power distance include:** Arab countries, India, Malaysia, Mexico, Philippines
- **Countries with lower power distance include:** Austria, Denmark, Germany, Ireland, Israel, Sweden, Switzerland, United States

2.4.2 Individualism/Collectivism

If someone asked you to finish the sentence, "I am ________," what would you include in your response? If you're from an individualistic culture you're more likely to mention personal traits, but if you're from a collectivistic culture you're more likely to mention relationships and group memberships. The dimension of individualism/collectivism refers to the degree to which members of a culture view themselves to be *independent* or *interdependent.*

Hofstede (1980) describes members of individualistic societies as self-centered, competitive, calculative, pursuing their own goals, and having a low need for dependency upon others and a low level of loyalty for the organizations. By contrast, he describes members of the collectivist societies as having a "we" rather than an "I" orientation, interacting with each other in an interdependent mode, and taking action jointly as a group in a co-operative fashion, subscribing to the moralistic values of joint efforts and group rewards, and having a high level of loyalty for the organization.

However, many factors can influence individualism/collectivism, so individuals within a culture can also differ in their levels of independence/interdependence. Individualism and collectivism can even be affected by the situational context. In Hofstede's research, this cultural dimension was shown to have a strong correlation with power distance. That is, individualistic cultures tend to have a preference for lower power distance and collectivistic cultures tend to have a preference for higher power distance (Figure 2.5).

- **Countries with more individualistic cultures include:** Australia, France, Italy, Netherlands, United Kingdom, United States
- **Countries with more collectivistic cultures include:** China, Malaysia, Philippines, Portugal, South Korea, Thailand, Venezuela

2.4.3 Uncertainty Avoidance

How comfortable, or uncomfortable, do you feel in unstructured situations? Unstructured situations are unknown and sometimes surprising. Uncertainty-avoiding cultures try to minimize the possibility of such situations. But *uncertainty* avoidance is not the same as *risk* avoidance. It has to do with the degree to

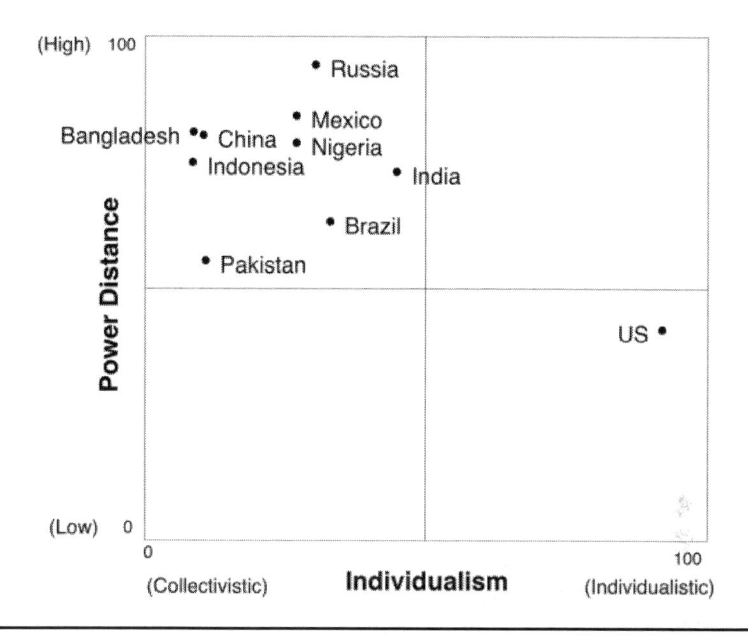

Figure 2.5 Cultural Profiles for the 10 Most Populated Countries Based on Hofstede's Power Distance and Individualism Cultural Dimensions (Reprinted from Geert Hofstede, Gert Jan Hofstede, Michael Minkov, *Cultures and Organizations, Software of the Mind,* Third Revised Edition, McGrawHill 2010, ISBN 0-07-166418-1. ©Geert Hofstede B.V. quoted with permission)

which members of a culture are willing to accept and deal with ambiguous situations (see Sidebar "Uncertainty Avoidance in Police–Civilian Interactions"). Cultures with strong uncertainty avoidance prefer structure and predictability, and this generally results in strict laws and explicit rules of behavior. Members of strong uncertainty-avoidance societies do tend to be risk averse when it comes to changing employers, embracing new approaches, or engaging in entrepreneurial activities. People are inherently more anxious about the unpredictability of the future than people in other cultures, and they perceive change as dangerous (Steensma, Marino, and Dickson, 2000). By contrast, members of weak uncertainty-avoidance cultures are more comfortable with unstructured situations and ambiguity, and this favors risk taking, innovation, and the acceptance of different views.

- **Countries with higher uncertainty avoidance include:** Austria, Brazil, France, Germany, Italy, Japan, Saudi Arabia, United Arab Emirates
- **Countries with lower uncertainty avoidance include:** China, Hong Kong, India, Philippines, Singapore, South Africa, United Kingdom

Uncertainty Avoidance in Police–Civilian Interaction

Researchers examined how the cultural dimension of uncertainty avoidance—a person's tolerance or intolerance for uncertain or unknown situations—impacts communication alignment in crisis negotiations. Using transcriptions of 53 negotiations from a Dutch–German police training initiative, in which police negotiators interacted with German (high uncertainty avoidance) and Dutch (low uncertainty avoidance) mock perpetrators. They found that formal language and messages, which emphasize laws and regulations, appear to be effective when addressing a perpetrator from a high uncertainty avoidance country (e.g., Germany), because these individuals are less tolerant of unknown or uncertain situations. However, they found that using this approach with perpetrators originating from a low uncertainty avoidance country (e.g., the Netherlands) is less effective.

These research findings show the effects of cultural background on communication and demonstrate the benefits of using more formal language and messages that emphasize law and regulations when interacting with people from high uncertainty avoidance cultures.

What are the implications of these findings for change management?

Source: Giebels et al. (2017)

2.4.4 Masculinity/Femininity

Do you live to work or work to live? Would you prefer to work the same hours and make more money or to work fewer hours and make the same money? If you're from a masculine society you are more likely to prefer the "more money" option, but if you're from a feminine society you are more likely to prefer the "fewer hours" option.

According to Hofstede, the masculine/feminine dimension has to do with how gender roles are distributed. Masculine cultures tend to reflect a dominance of tough values that are almost universally associated with male roles—such as competition, assertiveness, and material success. In masculine societies, both men and women tend to be assertive and competitive; however, women tend to be less so than men. Societies in which there is not a strong differentiation between genders for emotional and social roles are thought of as being feminine. Feminine cultures tend to focus on tender values such as quality of life, personal relationships, and care for others. In feminine cultures gender roles overlap, whereas they are clearly distinct in masculine cultures.

Masculine cultures are about ego. Feminine cultures are about relationships. In a business setting, the masculine/feminine continuum produces important differences in work content and management styles.

Feminine cultures tend to place a stronger emphasis on employee well-being and work–family balance, whereas masculine cultures place more emphasis on bottom-line performance, and work prevails over family. (Interestingly, but perhaps not surprisingly, Hofstede's [1980] IBM studies revealed that women's values differ less among societies than men's values.)

- **Countries with more masculine cultures include:** China, Germany, Hungary, Italy, Japan, Mexico, Switzerland, Venezuela
- **Countries with more feminine cultures include:** Denmark, Finland, Netherlands, Sweden, Thailand

2.4.5 Long-Term Orientation/Short-Term Orientation

Every culture maintains some connection to historic events and traditions while also facing present-day challenges and preparing for the future. Long-term versus short-term orientation refers to whether a society exhibits a pragmatic future-oriented perspective or a conventional historic point of view. In other words, it is a measure of a culture's perspective of how the future is felt to impact life and business, and how life and business impact the long-term view of the culture.

Cultures with a long-term orientation tend to emphasize virtues directed toward the future—perseverance, thrift, ordering relationships by status, and having a sense of shame. Short-term orientation cultures tend to foster virtues related to the past and present—respect for tradition, preservation of "face," and personal steadiness and stability.

Businesses in long-term-oriented cultures are generally accustomed to working toward building strong positions in their markets and do not necessarily expect immediate results. In short-term-oriented cultures control systems are focused on the "bottom line," and managers are constantly judged by it. For example, in Western cultures, time is a commodity. If you're not early, you're late. Time is money. But in two-thirds of the world, time happens when it's supposed to. It is much more flexible and elastic.

- **Countries with long-term orientation include:** Brazil, China, India, Japan, South Korea, Thailand
- **Countries with short-term orientation include:** Germany, New Zealand, Norway, United Kingdom, United States

2.4.6 Indulgence/Restraint

This dimension is more or less complementary to the long-term versus short-term orientation dimension (Figure 2.6). It has to do with the extent to which a society allows relatively free gratification of basic and natural human desires related to enjoying life and having fun (Hofstede, 2011). There is no absolute standard for the degree of indulgence. What is measured here is the position of societies relative to each other (Figure 2.6). Indulgent cultures tend to focus more on individual happiness and wellbeing. More importance is placed on freedom of speech and personal control. In restrained cultures there tends to be a greater sense of helplessness about personal destiny. Positive emotions are less likely to be freely expressed, and happiness, freedom, and leisure time are not given the same importance as they are in indulgent societies. In the workplace, this dimension is likely to have an impact on how willing employees are to voice opinions and give feedback. And in more indulgent cultures, employees may be more likely to leave an organization when they are not happy in their roles.

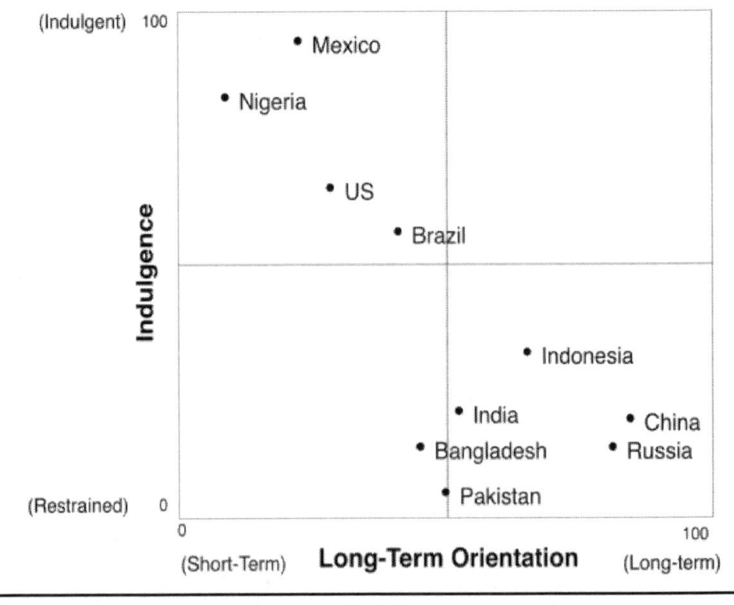

Figure 2.6 Cultural Profiles for the 10 Most Populated Countries Based on Hofstede's Long-Term/Short-Term Orientation and Indulgence/Restraint Cultural Dimensions (Reprinted from Geert Hofstede, Gert Jan Hofstede, Michael Minkov, *Cultures and Organizations, Software of the Mind,* Third Revised Edition, McGraw Hill 2010, ISBN 0-07-166418-1. © Geert Hofstede B.V. quoted with permission)

Differences in attitudes around customer service are another interesting aspect of the indulgence/restraint dimension. In indulgent cultures, customer service representatives are more likely to visibly demonstrate their "happiness" with a smile and friendly demeanor. In more restrained cultures, this is likely to be considered inappropriate and unnatural. US retailer Wal-Mart's failure in Germany is a good example (see Sidebar "Wal-Mart's Downfall in Germany.") Staff smiling at customers in the US is normal and expected, but in Germany customers found this practice unsettling. On a scale of 0 to 100, the average score for indulgence for the US is 68, but for Germany it is 40.

- **Countries with higher indulgent cultures include:** Argentina, Australia, Canada, Colombia, Iceland, Nigeria, Sweden
- **Countries with higher restrained cultures include:** China, India, Indonesia, Morocco, Pakistan, Russia

Wal-Mart's Downfall in Germany

In 1998, Wal-Mart moved into Germany hoping to repeat its phenomenal US success in Europe's biggest economy. It didn't turn out that way, and in 2006, after eight loss-making years and in a humbling admission of defeat, Wal-Mart bid *auf wiedersshen* to Germany. What went wrong?

No one can say precisely why the venture failed. From the time it entered the German market, Wal-Mart struggled against stiff competition from discount retailers already operating in Germany. But some cross-cultural peculiarities have also been identified as determining factors. From the beginning, Wal-Mart's American managers pressured German executives to enforce American-style management practices in the workplace. But the American approach to business did not quite translate into German.

For example, some of the mandated rituals intended to boost morale and instill loyalty—such as group chanting and stretching exercises at the start of each shift—did not go over well with the Germans. Nor did the hearty greetings from staff or flashing smiles at customers after bagging their purchase. Because Germans don't normally smile at complete strangers, the spectacle of grinning employees unnerved customers.

Wal-Mart's belief that it could apply its American success formula in an unmodified manner to the German market turned out to be a fiasco. And, interestingly, around the same time it retreated from Germany, Wal-Mart also withdrew from South Korea for similar reasons. Culture matters.

What change management lessons can we learn from Wal-Mart's failed ventures?

Sources: Barbara (2007), Knorr and Arndt (2004)

Over the years, Hofstede's research has served as a marker for subsequent researchers and continues to be widely cited. Although it has sometimes been criticized for its limitations—such as old data, one-company approach, and too few dimensions—there can be little doubt that Hofstede's model remains one of the most valuable pieces of work in the field of intercultural research. It continues to help organizations to understand how they can collaborate more effectively across cultures. By drawing on Hostede's dimension measures of different groups, you can begin to raise your own awareness and develop a broader understanding of the cultural differences between these groups relative to how they react to the work environment and change. For example, the combination of high power distance, collectivism, and high uncertainty avoidance may increase resistance to change, whereas the combination of individualism, low power distance, and low uncertainty avoidance may increase openness to the global world (Harzing and Hofstede, 1996). How valuable would this kind of insight be when you are planning change management interventions in different cultural contexts?

2.5 GLOBE Model

In one of the most ambitious and extensive cross-cultural research projects, an international team of researchers (led by Robert House of the Wharton School of Business at the University of Pennsylvania) focused on cultural differences in leadership. Titled the *GLOBE* (Global Leadership and Organizational Behavior Effectiveness) *Study of 62 Cultures* (House et al., 2004), the research empirically developed nine cultural dimensions to measure differences and similarities in norms, values, beliefs, practices, and leadership effectiveness across cultures based on data collected from 17,300 middle managers from 951 companies in 62 countries (Figure 2.7). Six of the GLOBE culture dimensions have their origins in the dimensions of culture identified by Hofstede: *uncertainty avoidance, power distance, institutional collectivism, in-group collectivism, gender egalitarianism*, and *assertiveness*. The other three dimensions drew on the work of other researchers (House et al., 2004). *Future orientation* and *humane orientation* had their origins in Kluckholm and Strodtbeck's (1961) *Past, Present, Future Orientation* and *Human Nature as Good vs Human Nature as Bad* dimensions, respectively. *Performance orientation* was derived from McClelland et al.'s (1953) work on the need for achievement.

The GLOBE study was designed to replicate and expand on Hofstede's work and to test various hypotheses that had been developed, in particular, on leadership topics. It included two forms of questions for each of the nine dimensions: one measured managerial reports of actual practices and values within the organization, and the other measured managerial reports of practices and values in their respective societies.

Power Distance	Degree to which members of a collective expect power to be distributed
Uncertainty Avoidance	Extent to which a society, organization, or group relies on social norms, rules, and procedures to alleviate unpredictability of future events
Humane Orientation	Degree to which a collective encourages and rewards individuals for being fair, altruistic, generous, caring, and kind to others
Collectivism I (Institutional)	Degree to which organizational and societal institutional practices encourage and reward distribution of resources and collective action
Collectivism II (In-group)	Degree to which individuals express pride, loyalty, and cohesiveness in their organizations or families.
Assertiveness	Degree to which individuals are assertive, confrontational, and aggressive in their relationships with others
Gender Egalitarianism	Degree to which a collective minimizes gender inequality.
Future Orientation	Extent to which individuals engage in future-oriented behaviors such as delaying gratification, planning, and investing in the future
Performance Orientation	Degree to which individuals are assertive, confrontational, and aggressive in their relationships with others

Source: House et al., (2004)

Figure 2.7 GLOBE Cultural Dimensions

After empirically establishing the nine cultural dimensions to capture cultural differences and similarities across societies, GLOBE was able to group countries into country clusters (Figure 2.8). Here, cultural similarity is greatest among societies that constitute a cluster, and cultural difference increases the farther clusters are apart. For example, in Europe, the Nordic cluster is most dissimilar from the Eastern European cluster.

GLOBE also analyzed survey responses to questions about leadership characteristics—integrity, modesty, decisiveness, etc. This generated 21 leadership scales, which were conceptually reduced to six leadership styles:

- **Charismatic/value-based.** Seeks to inspire people around a vision; creates passion among them to perform.
- **Team-oriented.** Highly values team cohesiveness and a common purpose or goals.
- **Participative.** Encourages input from others in decision making; emphasizes delegation and equality.
- **Humane.** Patient, supportive, and concerned with the well-being of others.
- **Autonomous.** Independent, individualistic, and self-centric approach to leadership.
- **Self-protective (and group-protective).** Emphasizes procedural, status-conscious, and "face-saving" behaviors; focuses on the safety and security of the individual and the group.

Anglo	Germanic Europe	Latin Europe	Eastern Europe	Nordic Europe
Australia Canada England Ireland New Zealand South Africa U.S.	Austria Germany Netherlands Switzerland	France Israel Italy Portugal Spain Switzerland	Albania Georgia Greece Hungary Kazakhstan Poland Russia Slovenia	Denmark Finland Sweden
Latin America	**Sub-Saharan Africa**	**Middle East**	**Southern Asia**	**Confucian Asia**
Argentina Bolivia Brazil Colombia Costa Rica Ecuador El Salvador Guatemala Venezuela	Namibia Nigeria South Africa Zambia Zimbabwe	Egypt Kuwait Morocco Qatar Turkey	India Indonesia Iran Malaysia Philippines Thailand	China Hong Kong Japan Singapore South Korea Taiwan

Source: House et al., (2004)

Figure 2.8 GLOBE Culture Clusters (adapted from source)

GLOBE grouped the culture clusters (Figure 2.8) according to the degree (higher or lower) to which they prefer each of the six leader styles (Figure 2.9). Countries grouped together do not differ significantly from one each other, but they do differ from grouped clusters at other points along the spectrum.

GLOBE's major finding is that leader effectiveness is contextual. That is, there is a correlation between national culture and leadership. "Leaders behave in a

	Charismatic	Team-oriented	Participative	Humane	Autonomous	Self-protective
Higher	Anglo Germanic Nordic SE Asian L. European L. American	SE Asian Confucian L. American E. European African L. European Nordic Anglo Mid. Eastern Germanic	Germanic Anglo Nordic	SE Asian Anglo African Confucian	Germanic E. European Confucian Nordic SE Asian Anglo African Mid. Eastern L. European L. American	Mid. Eastern Confucian SE Asian L. American E. European
	Confucian African E. European		L. European L. American African	Germanic Mid. Eastern L. American E. European		African L. European
Lower	Mid. Eastern		E. European SE Asian Confucian Mid. Eastern	L. European Nordic		Anglo Germanic Nordic

Source: House (2004)

Figure 2.9 GLOBE Culture Clusters and Leadership Styles

manner consistent with the desired leadership found in that culture" (Dorfman et al., 2012). Given that leadership/sponsorship is a key factor in both successful and failed change initiatives, taking into account the influence national culture has on leader behavior is critical.

2.6 Other Dimensions of Culture

American anthropologist Edward T. Hall identified three other important dimensions of culture: *context, time,* and *space.* Context has to do with the context in which cultural interactions take place. In high-context cultures, meaning is conveyed through context more than the message itself. In low-context cultures, messages are much more explicit, and the context is less important. North American and many European cultures tend to be low context, whereas many Middle Eastern and Asian cultures tend to be high context. But there are many different degrees between high and low context (Hall, 1959).

Hall categorized cultures into high-context cultures (where the communication style in which most of the information is already shared by people in the society, leaving very little information in the explicit transmitted part of the message) and low-context cultures (where the communication style in which most of the information is incorporated into the message and detailed background information is needed in the interaction with others).

For example, some cultures value a more direct, explicit approach to communication versus a more indirect, implicit approach. Direct communicators are often frustrated by what they perceive to be obtuse, unclear forms of communication. And indirect communicators are offended by what appears to them as a blunt, rude style. Germanic and Nordic cultures tend to favor direct, or *low-context,* communication. In Asian and African cultures, indirect, or *high-context,* communication is the norm. Managers will not be able to effectively address conflict on multicultural teams without understanding direct versus indirect approaches to communication.

The way we relate to time is another dimension of culture. How we view and manage time depends on what our culture considers good time management. People in places such as Italy and Brazil experience time differently from people in Switzerland and Germany. (See Sidebar "Running to Tico Time.")

Hall distinguished between *monochronic* and *polychronic* time orientations. People in monochronic cultures compartmentalize time. They tend to do one thing at a time, keep to a strict schedule, and are intolerant of lateness or interruptions. Northern Europe, Canada, and the United States are examples of societies with a monochromic time orientation. By contrast, people in polychronic cultures view time as cyclical. They typically keep several options running at once, and they tend to manage interruptions well with a willingness to change plans easily. People, not tasks, are their main concern. For example, promptness is based on the relationship rather than the task. Latin America, Arab countries, and sub-Saharan African countries have a polychronic time orientation.

In addition to context and time orientation, Hall also identified personal space as an important cultural dimension. The way we use space is as an elaboration of culture, and different cultures maintain different standards of personal space. How close we stand to our colleagues, our friends, or strangers varies widely among countries. Our personal boundaries have a lot to do with where we grew up. The world is divided into *contact* cultures and *non-contact* cultures. In non-contact cultures, such as Northern Europe and North America, people tend to stand farther apart and touch less than people in contact cultures, such as Southern Europe and Latin America. Recognizing cultural differences in the use of space can improve your cross-cultural understanding and help you better cope with any discomfort you may experience if the interpersonal distance is too large or too small based on your own standards of personal space.

Running to "Tico Time"

"I thought we were starting the meeting at 10:00, but it looks like everyone is running to Tico time."

Costa Ricans (Ticos) have a rather nonchalant attitude about time. "Tico time" is a reference to their habit of arriving late for appointments, even business meetings. This practice can be annoying to North Americans, Germans, and people from other cultures who are used to being precisely on time, if not early, for appointments.

Time is seen in a different light in different societies. The British, Americans, and Germans sanctify timekeeping. Time is wasted if tasks aren't being performed or decisions aren't being made. Time is money. But like the Ticos, Spaniards, Italians, Arabs, and others may ignore the passing of time if it means that conversations will be left unfinished. For them, completing a human transaction is the best way they can invest their time.

Everyday global business activities, such as scheduling a meeting or planning a change initiative, can be affected by attitudes to time. When doing business in other countries, you should consider the different perceptions of time people might have. Take time to observe how your own culture and the culture of others react to time.

2.7 Culture Matters

Effectively managing change across cultures requires an understanding of the influences of both the internal and external environment of the organization. Poor cross-cultural awareness can have consequences—some comical, such as an Italian's reaction to a foreigner ordering a cappuccino after 11 A.M., others serious, such as Wal-Mart's failure in Germany. Many change initiatives require us to work in a multicultural environment, often without awareness of the impact of cultural factors on the outcome of the project. The cross-cultural research can help you develop a macro understanding or cultural-general understanding of cultural systems, norms, and values, which may enable you to interact more effectively with leaders, managers, and employees from different backgrounds. For example, understanding the basic cultural norms for how men and women should relate or the culture's relationship to time might help you avoid being perceived as rude or possibly even inappropriate in your cross-cultural interactions. A simple starting point would be to review the cultural models developed by Hofstede, GLOBE, Hall, and others.

In addition to developing a cultural-general understanding, you may also need to develop a context-specific understanding, which includes understanding the relevance of culture to specific environments. For example, the way a technology company works across borders differs from the way a financial services organization does. Each requires specialized, domain-specific cultural knowledge in addition to a macro level understanding of overarching cultural differences.

Although you need to recognize that cultural differences exist, you also need to take care not to assign values—better or worse, right or wrong—to those differences. The term *sophisticated stereotyping* has been used to describe cultural dimensions research (Osland et al., 2000), but stereotyping can have a negative connotation. Let's think of the cultural dimensions as broad tendencies or norms rather than stereotypes. Of course, there is enormous diversity in all cultures, so it is dangerous to generalize about everyone from a specific culture or nationality. You cannot assume that all Germans are direct and all Chinese are hierarchical or that all Latin Americans are collectivists. But the cultural dimensions described above can give you the ability to identify when and how certain cultural values influence the way an individual from another culture may think, behave, or react to change. For example, participative management can improve performance in low power distance cultures but may worsen it in high power distance cultures, and emphasizing individual contributions may improve performance in more individualistic cultures but have the opposite effect in more collectivistic cultures (Newman and Nollen, 1996).

And the cross-cultural research can really help reveal how unique your own culture is, and how that can impact the way you do business with people from other cultural backgrounds. How do you think you can use your understanding of national culture to know when and how to develop culturally mindful change management interventions?

It is crucial for change leaders and change managers to understand the impact of cross-cultural differences on business and the organization. The success or failure of a change is essentially in the hands of people. If these people are not cross-culturally aware then misunderstandings, offence and a breakdown in communication can occur.

Key Points

- Culture can be defined as a shared meaning system that influences how we think, act, organize, relate, and perceive.
- Culture is a multi-layered, nested structure.
- Organizational culture is influenced by the national culture in which the company operates.
- Some elements of culture are easily observable (language, food, etc.), but others are invisible and more difficult to decipher (values, basic assumptions, etc.)
- Cultural differences can have an important effect on perceptions and reactions to change and need to be taken into account when leading and managing change across cultural contexts.
- Hofstede, GLOBE, Hall, and other researchers have constructed useful frameworks for understanding broad differences between national cultures, which can be leveraged when developing strategies for leading and managing change in cross-cultural environments.
- Differences in organization, leadership, and communication can be used to measure differences in groups and individuals and help you anticipate when and why cultures may clash.
- Cultural dimensions (norms) are high-level tendencies, but there is enormous diversity in cultures, so it can be dangerous to generalize about everyone from a specific culture or nationality.

Want to Know More?

Hofstede Insights (www.hofstede-insights.com) is a good place to start if you want to learn more about Hofstede's 6-Dimension Model of national culture. The site also includes a country comparison tool as well as news, blogs, and information on training and consulting services.

Geert Hofstede has written numerous books. You might be interested in reading *Cultures and Organizations: Software for the Mind,* Third Edition, (2005) or *Culture's Consequences: Comparing Values, Behaviors, Institutions, and Organizations Across Cultures* (2001).

GLOBE is an organization dedicated to the international study of relationships among societal culture, leadership, and organizational practices. You can learn more about GLOBE on its website, www.globeproject.com.

To learn more about Hall's ground-breaking research on the cultural dimensions of context, time, and space you might be interested in reading *The Silent Language* (1959) and *Beyond Culture* (1976).

Chapter 3

The West and the Rest

The top business schools are WEIRD, the world of management is WEIRD, and you are probably WEIRD too.[*]

— Dr. Leandro Herrero

The best product launches spend months beforehand understanding the consumer's needs, desires, and habits within the relevant markets. Shouldn't organizations consider taking this same purposeful approach with their cross-cultural change initiatives? And shouldn't we, as change management professionals, counteract our own potentially flawed assumptions by taking the time beforehand to understand how employees' beliefs and behaviors in different cultural contexts may influence their reactions to change?

Management theories—including the prevailing change management theories—make certain assumptions about human nature, and for several decades now a number of management researchers and social scientists have been questioning the applicability of Western management theory in non-Western contexts (e.g., Hofstede, 1980; Laurent, 1986). Western management theories and practices reflect the cultural environment in which they were written (Hofstede, 1980), and they cannot be separated from Western cultural tendencies toward individualism, lower power distance structures, uncertainty acceptance, and short-term orientation (see Table 3.1). Think about it. The very concept of "What's

* https://leandroherrero.com/weird-western-educated-industrialised-rich-and-democratic-its-our-management-education-crafted-in-this-model/

in it for me?" (WIIFM), for example, is steeped in Western cultural tendencies toward individualism.

Table 3.1 Western Cultural Perspectives Reflected in Management Practices

Leadership style	More participative, less hierarchical; use of incentives for economic advancement
Strategic focus	Preference for fast, measurable payback
Evaluation/promotion	Based on individual performance and contributions
Motivational systems	Greater emphasis on individual achievement, competition, and personal goals; desire for plans that focus on rewards for individual contributions; greater emphasis on quality of life/work-life balance
Remuneration	Extrinsic rewards based on market value; pay for performance
Decision making	Decentralized; participative; individual responsibility for decisions; designed for logical analysis of problems
Cooperation/partnership	Greater emphasis on contractual safeguards to ensure that partners' tendencies to focus on personal goals/aspirations do not interfere with one's own goals/aspirations
Work groups/teams	Higher importance placed on the task and building confidence for superior performance; less emphasis on team social and interpersonal relations (i.e., establishing personal relationships is not necessarily required for getting the job done)
Conflict resolution	Tendency to confront problems directly and bring things out into the open; preference for tactics that involve rational arguments, factual evidence, and suggested solutions

The concept of Western culture is generally linked to the classical definition of the Western world. The term, however, is not restricted to Western Europe but is also applicable to many countries whose history is strongly marked by Western European immigration or settlement (United States, Canada, Australia, New Zealand, South Africa, etc.).

Did you know that Western habits and cultural preferences are different from those of the rest of the world? Cultural psychologists have shown that people raised in "Western-educated industrialized rich democratic" (WEIRD) countries often exhibit different psychological processing than do people from less WEIRD countries. The very way Westerners think about themselves and others makes them distinct from other humans. Even the way Westerners perceive reality is different from how everyone else does. Some of the most notable differences are around the concepts of individualism (value personal success over group success) and collectivism (value group over the individual). Generally speaking, WEIRD populations are socially oriented to think of themselves as individuals entitled to free expression, even if that means violating social norms or acting in ways that breach traditional expectations. WEIRD societies actually account for only about 12 percent of the world's population, and they tend to be outliers in the way they perceive and react to the world around them (Henrich, Heine, and Norensayan, 2010). In other words, the WEIRD are the weird. Ironically, the world of business management—and this includes change management—is dominated by Western-oriented theory and practices, and this has led to the belief that effective Western management practices can be applied uniformly and with equal effectiveness across geographies and cultures. We really need to question this thinking.

As you have already learned from the previous chapter—and perhaps from firsthand experience—national culture can influence our understanding of work, our approach to it, and the way in which we expect to be treated in the workplace. This implies that one way of acting or one set of outcomes is preferable to another, depending on the cultural context (see Table 3.2).

For example, many organizations in WEIRD societies adopt practices such as information sharing with employees and participative decision making based on the assumption that employees are comfortable with and desire increased empowerment, autonomy, and participation. This assumption reflects the cultural norms of mostly-WEIRD societies, but the extent to which these practices

Table 3.2 Cross-Cultural Orientations: Western versus Non-Western Societies

Western (WEIRD) Tendencies	Non-Western (non-WEIRD) Tendencies
Individualistic orientation; inner bias of "me"	Collectivistic/group orientation; inner bias of "we"
Need for self-assertion	Need for coordination
Attribution groups are important (immediate family, class, occupation)	Frame groups are important (village, neighborhood, company, region, nation)
Contractual relationships (based on rights and duties)	Personal relationships (based on mutual obligations and mutual dependence)
Behavior controlled by rules, punishments, and rewards	Behavior controlled by group adaptation (violation of group norms results in feelings of shame)
Weak hierarchical structure	Strong hierarchical structure
Importance placed on freedom and personal conscience	Importance placed on security and obedience
Emphasis on absolute moral values (good versus evil)	Emphasis on virtue

work in more hierarchical, less-WEIRD societies is debatable. Similarly, the notion that employees are a strategic source of competitive advantage that creates value for organizations is generally accepted in WEIRD business management models, as opposed to management approaches in which employees are viewed more as tangible resources to be obtained cheaply, controlled, and exploited as fully as possible. Values and behaviors do differ across cultures, and differences in culture call for differences in management practices. When management practices are inconsistent with the culture, employees are more likely to feel dissatisfied, uncommitted, and resistant to change.

Of course, WEIRD or not, you can't assume that everyone from a given culture will react to a situation in exactly the same way. In spite of the norms of the culture to which we belong, at times we may exhibit different behaviors in our different environments. Think, for example, about someone from China working locally (in their home country) for an American company. The culture of the organization may be strongly influenced by Western social norms that may be quite different from the local culture. As a result, the Chinese employee may act more individualistically in the office, while their cultural heritage may foster strong collectivism in other relationships. Context matters.

3.1 Is Change Management too WEIRD?

Probably. Like other business management theories, the dominant change management models, frameworks, standards, bodies of knowledge, and so on have largely been developed in WEIRD countries and, like many other business management models, are imbued with Western management theory, social orientations, and assumptions. Now, this may not be of consequence if you are WEIRD and only work in WEIRD countries, but if you are working in non-WEIRD countries, you may find that the WEIRD way of doing things does not work well in every cultural context, and that what made you a successful change agent in a domestic or local context may not help you reach the same level of success on a global scale.

The content of change management is reasonably correct, but our capacity to implement it in different cultural contexts—both Western and non-Western—has been woefully underdeveloped, and the consequences of getting things wrong can be dramatic. As globalization accelerates, organizations will be increasingly exposed to unfamiliar cultural contexts and diverse workforces. We will need global approaches with different emphasis and elements suited to the encountered cultures and new global skills to achieve effective interaction and change interventions in multicultural contexts.

Reflect on your own attitudes toward work. What aspects of your cultural, family, and professional background influenced your attitude, expectations, and beliefs about work? Now, think about how different countries or regions compare in terms of embracing or resisting new trends in work (e.g., change). What would you take into consideration when planning interventions to address attitudes toward work and change in different cultural contexts?

Key Points

- Business management is dominated by Western-oriented theory and practices, and this has led to the belief that effective Western management practices can be applied uniformly and effectively everywhere.
- Western management theories and practices reflect the cultural environment in which they were written, and they cannot be separated from Western cultural tendencies.
- WEIRD (Western Educated Industrial Rich Democratic) societies actually account for only about 12% of the world's population, and they tend to be outliers in the way they perceive and react to the world around them.
- The dominant change management models, frameworks, standards, and bodies of knowledge have largely been developed in WEIRD countries and, like many business management models, are imbued with Western management theory, social orientations, and assumptions.
- Global approaches embrace different emphases and elements suited to the encountered cultures and new global skills to achieve effective interaction and change interventions in multicultural contexts.

Want to Know More?

To learn more about WEIRD theory, you can watch Joseph Henrich's thematic series *The Emerging Science of Culture: The Weirdest People in the World* on YouTube at https://www.youtube.com/watch?v=dmlfPY4T9bk

You might also be interested in reading *The Weirdest People in the World?* by Joseph Henrich, Steven Heine, and Ara Norenzayan. Their paper originally appeared in 2010 in the prestigious journal *Behavioral and Brain Sciences* (Vol. 33), but a PDF version is available online at http://hci.ucsd.edu/102b/readings/WeirdestPeople.pdf

Part II

Culturally Tuning Change Management: Putting Cultural Research into Practice

*Knowledge is of no value unless you put it into practice.**

— Anton Chekhov

A few years ago, the Canadian Center of Science and Education published a compelling case study relating to the role of cultural intelligence during an enterprise resource planning (ERP) implementation in Thailand (Meissonier, Houzé, and Lapointe, 2014). The system itself was inconsistent with prevailing cultural values because it involved imposing "best practice" process designs based on a unique Western-oriented (WEIRD) managerial paradigm. As a result, it was imperative for the team to develop appropriate strategies for managing anticipated resistance from users facing the imposition of changes that did not align with the local culture and normal ways of working.

Thailand is a country in which managerial practices are generally specific and different from those in the United States or Europe, where ERP systems are

---

* https://www.brainyquote.com/quotes/anton_chekhov_119058

largely diffused across business sectors. It is also a culture steeped in Buddhism (95 percent of the population is Buddhist), and this influences how Thai people behave in both private and professional contexts. Furthermore, the relationship to time is more long-term oriented, meaning even the concept of "project" is considered in a culturally different way than it is in Western cultures. For example, "starting a project" does not mean that business transactions can start straight away just because a contract has been signed. A "trust period"—lasting several weeks or months—must be respected before resources and budgets can be established. This is different from a feasibility analysis and corresponds to "privileged moments," during which people get acquainted and have discussions not strictly related to the project.

In Thailand, face-to-face communication also plays an important role in building and maintaining the social legitimacy of hierarchical superiors. Nonverbal communication is used to express both good and bad feelings. For example, Thai people use facial expressions to express negative feelings without being forced to verbalize them. There are 13 different types of codified smiles expressing particular feelings (joy, sadness, annoyance, embarrassment, disagreement, etc.) that can intimate consent or dissent.

The team applied a cultural intelligence framework to better understand how observed cultural misalignments might translate into user reactions and how those reactions could be addressed so they did not become barriers to the change (adoption and usage of the ERP system). Taking into account the role of devotion and trust in the culture, the hierarchical cultural distance, and the tendency toward upward delegation, the change management approach took the form of an authoritarian management style.

Despite some dissatisfaction, users' acceptance of the ERP system was generally attributed to the consideration given by employees to top managers. This is akin to the "Bunkun" principle of Buddhism, which translates as "thankfulness" and corresponds to the gratitude of Buddhists toward those providing help.

This case study illustrates the importance of culturally tuning change management interventions. In this situation, the project team anticipated that there would be resistance to the implementation of the ERP system because the underlying processes were based on Western-style "best practices" that were inconsistent with the values of Thai culture. To deal with this, the team adopted an authoritarian approach that aligned with cultural expectations regarding leadership and decision making, in which respect for the power structure trumps employees' dissatisfaction with the change.

When an organizational change involves people with the same value orientations, business management processes tend to flow naturally and smoothly, and expectations are mutually understood. But when an organizational change involves individuals with different value orientations, there is a risk that the

A View of Thailand Through the Lens of Hofstede's 6-Dimensions Model

Thailand scores 64 on the Power Distance spectrum, which is slightly lower than the average for Asian countries (71). It is a society that observes a strict chain of command and protocol. The attitude toward managers is more formal, and the information flow tends to be hierarchical and controlled.

A score of 20 on the Individualism dimension means that Thailand is highly collectivistic. Thai society fosters strong relationships in which everyone takes responsibility for members of their in-group. Personal relationships are key to doing business in Thailand, but it can take time to build those relationships. A "trust period" is typically required before business transactions can begin.

Considered a feminine society, Thailand is less assertive and competitive than more masculine societies. With a score of 34, it ranks lower on the on the Masculinity dimension than other Asian countries and lower than the worldwide average of 50.

The Uncertainty Avoidance score for Thailand is 64. This high score suggests the society does not readily accept change. Change has to be seen as being for the greater good of the in-group.

Thailand's low score of 32 on the Long-Term Orientation dimension indicates that Thai culture is more normative than pragmatic. There is a great respect for tradition, but also a focus on achieving quick wins.

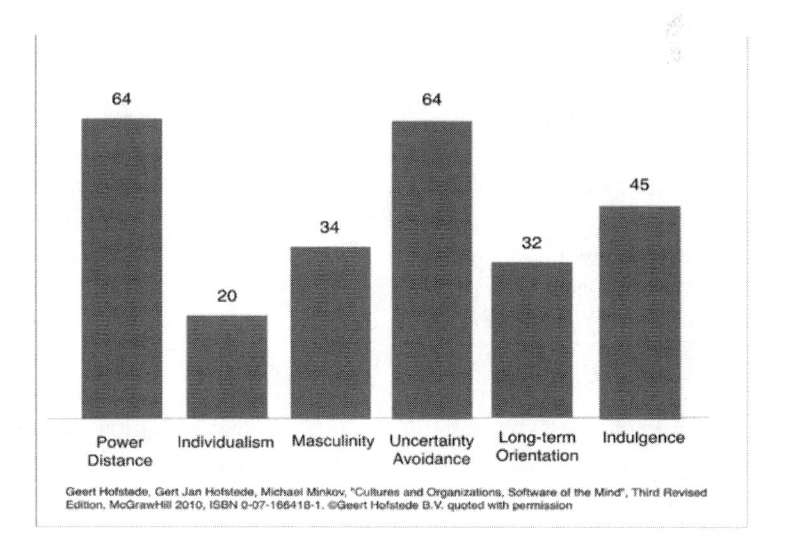

differences won't get acknowledged and the business management processes of the dominant group will be imposed. Managing change in culturally diverse settings can be tricky, and when using universal approaches to change, it is important to recognize that they might not serve their purpose in every cultural setting. We live in a global world, and every culture differs in terms of its approach to decision making, problem solving, motivation, and change. Today it's no longer enough to know how to lead and manage change the American way or the Japanese way, the Italian way or the Chinese way. You need to be informed enough and flexible enough to adapt your approach based on any particular cultural markers that could be potential barriers to (or catalysts for) effective change management (See Sidebar "National Culture and the Adoption of Emerging Technologies").

When working with enterprises from different cultures, you should consider whether transferring change management practices and processes from your own culture will be acceptable for employees in those other cultures. For example, when managing change outside your culture, transferring practices from your home country is more likely to be acceptable by your host country's employees if the host country culture is closer to (similar to) your home country culture. Some companies operating outside their cultures do not possess cultural characteristics of the cultural environment in which they are operating. This can create problems related to communication and understanding the expectation of the local employees and their working environment, among other things. In many cases, the management of the home culture assumes that their home country leadership approach will work well in the new and culturally different country. For this reason, you should have an idea how the different cultural dimensions can influence change management practices and activities, and you should have an idea about how distant or close your own culture is to the other cultures you encounter. Are they close and familiar or distant and exotic? The benefits of doing this are increased cultural awareness, decreased chance of derailment, and the opportunity to improve the working relationships of culturally diverse teams.

How can you use your understanding of national culture to lead and manage change more effectively? The short answer is that improving the alignment between your change management practices and cultural contexts can yield tangible results. Hofstede's cultural dimensions framework is a good starting point to expose and deepen your thinking about areas in which culture may impact your change management efforts and where you may need to apply culturally mindful facilitating strategies.

In the following chapters we'll take a closer look at the potential influence cultural characteristics can have on leadership and decision making, team dynamics, communication, learning, and resistance management, as these are some of the areas you will typically need to assess when planning and implementing change. Bear in mind that the dimensions represent a continuum, and most countries fall somewhere along the continuum rather than absolutely at one end or the other. And as Hofstede himself says, "It's important to remember that cultural dimensions don't exist in real life. They are only a way of understanding a very complex world. As such, the dimensions help us understand that what happens in one particular culture does not necessarily happen in another" (Pogosyan, 2017).

National Culture and the Adoption of Emerging Technologies

To what extent can national culture influence whether or not and why nations adopt new technologies? Which cultural variables help explain cross-country adoption rates of emerging technologies in corporate contexts?

Andrew Barron and Dirk Schneckenberg (2012) of the ESC Rennes School of Business in France proposed a model showing how some national cultural characteristics influence the adoption of emerging technologies, specifically Web 2.0 technologies. Their research assumed that technologies may enjoy faster adoption rates in organizations operating in countries where national cultures reject power distance, embrace collectivism, and accept uncertainty.

Using Hofstede's cultural dimensions, Barron and Schneckenberg looked at differences in power distance, individualism, and uncertainty avoidance across China, Finland, France, Norway, Spain, Sweden, and the United Kingdom.

Their findings suggest that, compared to their counterparts in other countries, the adoption and usage of Web 2.0 technologies will be high in and Finland, Norway, Sweden where national cultures exhibit uncertainty accepting tendencies, reject differences in power, status and authority, and are relatively collectivist in their cultural orientation.

If national culture can influence technology, what do organizations need to consider when implementing technological change initiatives across cultures?

Chapter 4

Adapting to Cultural Differences in Leadership Behaviors and Decision Making

Absorb what is useful. Discard what is not.
*Add what is uniquely your own.**

— Bruce Lee

Just as organizational culture is intertwined with leadership (Schein, 2010), cultural characteristics can also shape leadership perception and value judgments. And when it comes to decision making, the process can become more complex when there are two or more cultures involved, especially when the cultures involved are culturally distant (highly dissimilar). Beyond stereotypes or superficial generalizations, a cultural analysis of leadership and decision making can enhance your cross-cultural capacity and improve your chances for collaboration in different cultural contexts. Leaders who disregard culture, particularly during intercultural interactions, will adversely impact the effectiveness

* https://medium.com/bettertodayclub/absorb-what-is-useful-discard-what-is-not-add-what-is-uniquely-your-own-ad86e99fbd49

of their leadership (Hendrickson, 2016). Conversely, leaders who are culturally aware and behave accordingly are more likely to be effective (Du Plessis, 2011; Mustafa and Lines, 2012).

Becoming a global player requires cultural insight and an adaptable approach. Drawing on Hofstede's 6-Dimensions Model, Table 4.1 summarizes how each of the cultural dimensions can influence leadership and decision making and highlights some of the things you should consider when working with leaders in different cultural contexts.

Although Table 4.1 examines each cultural characteristic individually, it is important to recognize that there is a correlation between the cultural dimensions. For example, individualistic cultures tend to prefer low power distance structures, whereas collectivistic cultures show a greater preference for high power distance structures. Each combination of cultural variables can influence things like the speed of decision making, acceptance of unpopular decisions, level of commitment for implementing a decision, and reactions to change in very different ways. From this analysis, we can imagine that employees from collectivistic and high power distance cultures will have a higher degree of trust and confidence in executive management and a better understanding of and agreement with the company's vision than will employees from individualistic and low power distance cultures. This suggests that less time and effort is needed to "sell" the change to employees in collectivistic and high power distance cultures.

Naturally, these cultural values alone are not enough to explain leaders' behaviors without taking other contextual factors into account. Cultures are not monolithic and, within each societal culture, people can range along the continuum of each cultural dimension, no matter the overall country ranking (Hofstede, 2001). Differences in leadership style and decision making can also be attributed to differences in the personality of individuals and to other variables such as industry, profession, age, gender, and cultural intelligence.

How would you describe leadership behaviors and decision making in your own culture? What is your own perception of a good leader? When you are involved in cross-border change initiatives, it is essential that you understand your own leadership style and behaviors and how distant your own culture is from the culture(s) in which you are operating—that is, the overall degree of differences in key cultural dimensions between your own culture and the culture of the leaders and managers sponsoring and/or impacted by the change. This insight can help you better understand the management approaches that may await you in countries with different cultural backgrounds and to define the kind of approach you will need to apply.

Your aim should be learning to recognize and manage cultural distance effectively in both domestic and international organizational change assignments to improve your interactions with leaders in different cultural contexts. How will

(text continues on page 55)

Table 4.1 Influence of Culture on Leadership and Decision Making

Cultural Dimension	Things to Consider
Power Distance As it relates to leadership styles and behaviors, power distance deals with the expectations and distribution of power, authority, and status. Various studies have used this dimension to analyze the relationship between supervisors and subordinates and how they differ across cultures and subcultures. In low power distance cultures in which there is a limited dependence of subordinates on their supervisors, there tends to be a preference for a more democratic management approach. In high power distance cultures in which subordinates tend to have a stronger dependence on their supervisors, there is a preference for an autocratic and paternalistic management approach. In a paternalistic relationship, the role of the superior is to provide guidance, protection, and care to the subordinate, and the role of the subordinate, in return, is to be loyal and deferent to the superior. Although in Eastern cultures paternalism is one of the most desired characteristics of people in authority, it is generally viewed negatively in Western societies. In a Western cultural context, paternalism implies authoritarianism. In low power distance culture managers consider their subordinates' suggestions before taking any final decisions, whilst in high power distance cultures, only managers are involved in that process (Hofstede, 2001). Formal management systems typically dictate how strategic, functional, and tactical decisions are made and who needs to be involved in the decision-making process. Participative decision making tends to be more prominent in low power distance cultures, whereas a more non-participative approach tends to be more prominent within high power distance cultures.	Leaders in higher power distance cultures tend to adopt a more directive approach, whereas leaders in lower power distance cultures tend to adopt a more participative, supportive, and achievement-oriented approach. Ask yourself: Which style is most prevalent in your own culture? What do people think about their relationships with their leaders/managers and their subordinates? Is there is a large gap between them, or do leaders/managers expect subordinates to speak out? When using power distance as a measurement, it is important to also think about the different processes and structures that help shape leadership styles and behaviors. Power distance also seems to affect how long it takes to reach a decision. In a situation in which one party has less autonomy than the other, the decision-making process could take longer than anticipated. Based on your observations and experience with the organization, ask yourself: • Is the structure of the organization more hierarchical or more egalitarian? What are the implications for change sponsorship and access to sponsors? • What is the relationship between management and subordinates? Do leaders delegate authority to subordinates, or do they hold onto power? What are the implications for coalition building to support the change and communication? • Are influencing behaviors more implicit or more directive? What are the implications for communication, resistance management, and building support for the change? • What is the general degree of control held by senior managers, middle managers, and employees? What are the implications for

(continues on next page)

Table 4.1 Influence of Culture on Leadership and Decision Making (cont.)

Cultural Dimension	Things to Consider
Hofstede has placed a number of cultures on a continuum from low power distance to high power distance. Very few cultures are at one extreme or the other, and most cultures typically have a mix of high–low power distance characteristics, to some extent depending on where they fall on the continuum.	resistance management and reinforcing the change? • Is decision making more centralized or more participative? • Do employees take initiative, or do they tend to defer to the boss and wait for explicit instructions? What are the implications for the roles of middle managers and supervisors? • What level of involvement and freedom do employees have when it comes to decision making? Is it okay to disagree with the boss? • Are decisions made quickly or slowly? In order to get things done, is it acceptable to bypass the hierarchy, or must the chain of authority always be followed?
Individualism/Collectivism Some cultures emphasize personal responsibility and decision making, whereas others favor shared responsibilities and consensus building. Leadership style and decision making depends on what is appropriate and expected in each cultural environment (and within the organization). In individualistic cultures, leaders tend to be achievement oriented and more willing to take risks, resulting in an expansive-decisive strategy. By contrast, leaders with collectivist values tend to pay more attention to the social aspects of problems and are more sensitive to the social consequences of their actions. They tend to value security and follow passive, collaborative, and avoiding strategies. Hofstede has placed a number of country cultures on a continuum from Individualism to Collectivism. Very few cultures are at one extreme or the other, and most cultures typically have a mix of individualistic–collectivistic characteristics, to some extent depending on where they fall on the continuum.	In individualistic cultures, managers are more prone to adopt a transactional leadership style whilst collectivist cultures have shown a preference for transformational style. There is a also a correlation between individualism–collectivism and power distance. Collectivist cultures tend to favor high power distance (hierarchal) structures, whereas individualistic cultures tend to favor low power distance (egalitarian) structures. Based on your observations and experience with the organization, ask yourself: • Is the organization operating in a more individualistic or collectivistic culture? • Is the organization a foreign-owned enterprise? If so, how similar or different is the organization's home culture to the local culture in which it is operating? • What are employees' expectations of management? • Do leaders and managers consider their subordinates' suggestions before making final decisions?

(continues on next page)

Table 4.1 Influence of Culture on Leadership and Decision Making (cont.)

Cultural Dimension	Things to Consider
	• Are individual opinions or group opinions more important? • Is decision making more data oriented or more dialogue oriented? • Do employees expect to be included in decision making, or do they expect their leaders to be the decision makers?
Uncertainty Avoidance Some cultures are uncomfortable with ambiguity (unknown, unpredictable outcomes), whereas others emphasize flexibility and adaptability. High uncertainty avoidance cultures tend to be more rigid and intolerant of ideas that deviate from certain principles. They have a preference for clear, unambiguous, and formalized policies and procedures. By contrast, low uncertainty avoidance cultures have a preference for a limited number of policies and procedures. Managers from high uncertainty avoidance cultures are perceived to adopt behaviors which are more controlling, less delegating, and less approachable than managers from low uncertainty avoidance cultures (Offermann & Hellmann, 1997).	It is important to recognize that there is a difference between risk and uncertainty. Risky situations are those in which you know all alternatives, consequences, and their probabilities. Under conditions of uncertainty, however, not everything is known for sure. Researchers have shown a strong theoretical relationship between perceptions of uncertainty and aspects of decision making and policy formation (e.g., Hmieleski & Ensley, 2007), and uncertainty avoidance can also affect the amount of time decision making takes. Based on your observations and experience with the organization, ask yourself: • Are transformational leaders or controlling leaders more appealing in the culture in which the organization is operating? What are the implications for the specific change that is being introduced? • Are leaders willing or unwilling to say, "I don't know"? • Does decision making tend to be stable and cautious or intensive and flexible? • Is there a general tolerance or intolerance for the opinions of others? • Do decisions tend to be made quickly or slowly?
Masculinity/Femininity Some cultures place a premium on assertiveness, aggression, and toughness, whereas others value collaboration and collegial behavior. For example, Japan is a highly masculine culture (maximum emotional and	In feminine cultures, the approach to problem recognition is more subjective, whereas in masculine cultures it is more objective. But it is important to recognize that workers' general notions about the effectiveness of male and female managers can be as important as

(continues on next page)

Table 4.1 Influence of Culture on Leadership and Decision Making (cont.)

Cultural Dimension	Things to Consider
social role differentiation between genders), whereas Sweden is a highly feminine cultural (minimum emotional and social role differentiation between genders). Masculinity–femininity has been used to a lesser extent to analyze studies related to organizational leadership behavior and styles. One study (Adsit et al., 1997) found that in more feminine cultures, managers were more prone to display behaviors which emphasized cooperation and good working relationships, whereas in masculine cultures there was more emphasis on promoting an assertive, challenging, and highly ambitious working environment.	their actual leadership abilities or business results. Based on your observations and experience with the organization, ask yourself: • Are tough/aggressive leaders or caring leader/mediators more the norm in the culture in which the organization operates? • Is there a tendency in the culture to make a distinction between gender roles? What are the implications for leadership and decision making?
Long-Term/Short-Term Orientation Some cultures view adaption and circumstantial problem solving (pragmatism) as a necessity (long-term orientation), whereas others respect for tradition and stick to formal rules and procedures (short-term orientation). Long-term orientation tends to be higher in East Asian countries and lower in Latin American countries. China is very future oriented, whereas Mexico is more focused on the here and now.	People in societies with a short-term orientation tend to have a strong concern with establishing the absolute truth, whereas people in societies with a long-term orientation tend to believe that truth depends on the situation, time, and context. Based on your observations and experience with the organization, ask yourself: • What is the relative importance of here and now versus the future? Do leaders in the organization focus more on quick results or on persistence and long-term goals? • Is personal steadfastness and stability in leadership more important than personal adaptability? • Is there a strong desire to establish an absolute truth, or does truth depend on context? • Once decisions are made, are they flexible based on circumstances, or are they not easily adjusted once they have been made?
Indulgence/Restraint Cultures vary in the degree to which they stress conformity. Nordic, Anglo, and Latin American cultures value freedom of expression, whereas Eastern European, East Asian, and Muslim cultures emphasize conformance to social norms.	Based on your observations and experience with the organization, ask yourself: • Are leaders motivated more by gratitude and empowerment or material rewards for a job well done? • How important are status objects?

(continues on next page)

Table 4.1 Influence of Culture on Leadership and Decision Making (cont.)

Cultural Dimension	Things to Consider
Minkov (2011), who coined the label "indulgence versus restraint," provides an example of contrasting indulgent Americans and restrained Asians and Eastern Europeans. "Americans like to receive compliments. But in Japan and China, just like Eastern Europe, personal praise often causes embarrassment" (Minkov, 2011, p. 95).	• Do leaders tend to be more optimistic or pessimistic and cynical?

you bridge the differences in cultural characteristics? For example, if you are from a low power distance culture working in a high power distance environment, how will you need to adapt your approach in order to be more effective? Let's work through an example.

Imagine a team from Finland led by a woman has been tasked with helping a partner company in China prepare employees for a large organizational change. In preparation for the engagement, the Finnish team draws on Hofstede's cultural dimensions to assess how close or distant Chinese culture is to Finnish culture in terms of leadership behaviors, decision making, and employee expectations of management (Figure 4.1 and Table 4.2).

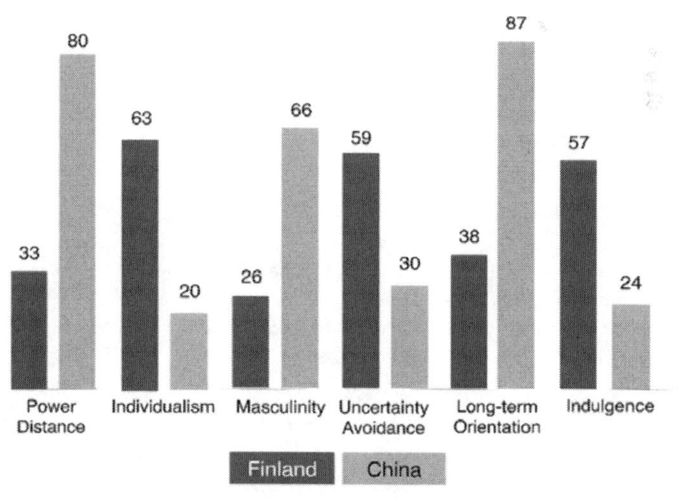

Figure 4.1 Cultural Profiles for Finland and China Based on Hofstede's 6-Dimensions Model (Geert Hofstede, Gert Jan Hofstede, Michael Minkov, *Cultures and Organizations, Software of the Mind,* Third Revised Edition, McGraw Hill 2010, ISBN 0-07-166418-1. © Geert Hofstede B.V. quoted with permission)

For additional insight, the Finnish team also takes into consideration the findings of the GLOBE study focusing particularly on cultural perceptions of what constitutes outstanding leadership (Figure 4.2 and Table 4.3 on page 59).

From this high-level comparison of the cultural profiles of Finland and China, the Finnish team can already see that there are significant differences in terms of power distance, individualism, uncertainty avoidance, future orientation, and gender egalitarianism. This analysis gives the Finnish manager insight into the cultural values that underpin Chinese values in the workplace and values that are considered necessary to create the best leadership style in a Chinese (Confucian Asia) context. This enables one to expect to be working with people who practice

Table 4.2 Comparison of Finland and China Based on Hofstede's Cultural Dimensions: Implications for Leadership Behaviors and Decision Making

Finnish Characteristics	Chinese Characteristics
• Low power distance • Individualistic • Uncertainty avoiding • Feminine • Short-term oriented (normative) • Indulgent	• High power distance • Collectivistic • Uncertainty accepting • Masculine • Long-term oriented (pragmatic) • Restrained
Leadership Profile	**Leadership Profile**
The Finnish management approach tends to be democratic. There is a lower distance (less inequality) between leaders and employees.	The Chinese management approach tends to be more autocratic. There is a higher distance (greater inequality) between leaders and employees.
Decision making is more decentralized, but Finnish leaders tend to bear alone the responsibility for the decisions they have made.	Decisions are made by people in authority, but responsibility for those decisions is then born by the entire group.
When Finnish leaders enter negotiations, they have certain limits. If the discussions go beyond these limits, there can be a process of consultation and debate among management before an actual decision is made.	Chinese bosses are expected to be arbitrary and act without explanation, but they may find it difficult to admit to a lack of knowledge or mistakes because this would cause them to "lose face."
Speaking up and making suggestions for improving operations is acceptable and expected. Subordinates are not afraid to challenge authority figures or say negative things out loud, even if it leads to conflicts.	Questioning authority figures is not acceptable. In order to prevent a manager from losing face, subordinates may shy away from making suggestions or may not inform that manager of problems.
The number of women in managerial positions is quite high. (There is a law which guarantees that at least 40 percent of committee and board posts in the public sector are held eitherby men or women.)	It is becoming more common for women to hold executive positions, but they may still find resistance to their leadership. (However, Chinese businessmen generally try to adjust their expectation of, and behavior toward, foreigners.)

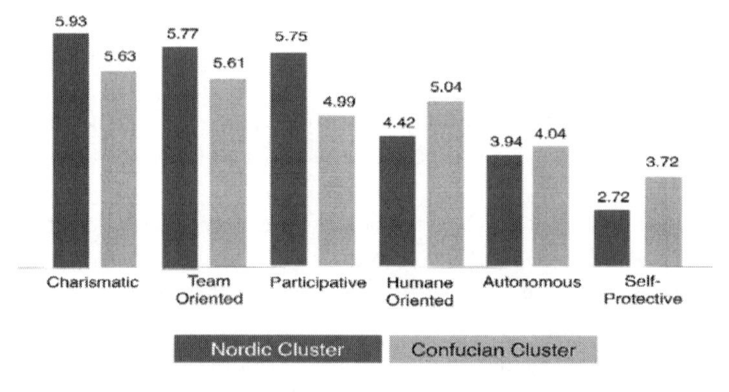

Figure 4.2 Leadership Profiles for Nordic Europe and Confucian Asia Culture Clusters Based on the GLOBE Study

much closer supervision than one is accustomed to. Most Finnish workers are strong initiative takers when compared to other cultures. They do not wait for instructions from the boss before starting a new task, and they expect to have enough responsibility to ensure their own success.

So, how should the Finnish manager adapt her style when working with Chinese leaders, peers, and subordinates?

Here are some of things she might need to do:

- Understand the principles of power distance. (In very high power distance cultures, such as China, the lower-level person will unfailingly defer to the higher-level person and feel relatively tolerant of it as the natural order. Leaders and employees have similar expectations regarding power distance, so there is general agreement as to who should be making the decisions.)
- Understand the chain of authority and its implications. Give respect to senior people by referring often to the next level up in the hierarchy.
- Don't jump straight into business. Understand the importance of establishing relationships of trust and respect with Chinese subordinates and management. A "trust period" may be required in before she can actually get down to business.
- Be ready to accept more direct orders from the sponsor and other more senior leaders that she might like.
- Suggest new ideas to the Chinese superiors first before trying them out. Change the minds of key influencers, and then leverage their relationship power to change others.

- Accept that Chinese subordinates may like strong supervision and feel comfortable with a directive, persuasive manager. But understand that directness needs to be done in a way that does not insult the individual.
- Convey a strong presence and decisiveness. (She may well have to be more authoritarian than she is used to, which means directions should be clear and explicit and deadlines should be stressed. She should not expect subordinates to take personal initiative unless they have been coached to do so.)
- Be aware that stating "I don't know" may not be an acceptable answer to subordinates who demand leadership from her, and she may lose their respect as a result.
- Understand that saving face is extremely important. Never put someone in a position where they have to disagree with their superior. (Subordinates may take care not to express their opinions in front of the boss until they learn what the boss thinks, because a disagreement could be viewed as finding fault with the boss.)
- Meet with a cross-cultural consultant to help her understand the Chinese mentality, values, and ways of working.

Now, the Finnish manager will not only need to consider the unique cultural perspectives of her Chinese superiors, peers, and subordinates, but also of her Finnish colleagues, as well as the cultural perspectives of any other key stakeholders from different cultural backgrounds. With an awareness of how culture can influence leadership behaviors and employee expectations about leadership, she can start to build culturally mindful strategies to encourage appropriate interactions.

If you were in this situation, how could you build on your knowledge of cultural differences around leadership behaviors and decision making to pull individuals together to successfully manage change? How could you use the differences to make a difference, in very practical ways, to the way you lead and manage change? Would gender be a factor in cross-cultural interactions?

The Finnish–Chinese example used here is, of course, oversimplified. In practice, there may be many other contextual variables to take into account. Among other things, the population of Finland is smaller and more homogenous than is the population of China, so generalizing Finnish culture is a bit easier than generalizing the culture of China with its large population, cultural pluralism, and diversity, in which values, beliefs, and conditions can vary from region to region. In practice, rather than thinking of China as a single culture, it may be necessary to take regional, ethnic, and subcultural differences into account. Even then, no two situations, no two change initiatives, will be exactly the same, and there may be many other contextual variables beyond national cultural characteristics that you would need to take into account (i.e., gender, age, industry, religious influences, etc.). Notwithstanding, there are some generally

Table 4.3 Comparison of Finland and China Based on GLOBE Cultural Dimensions and Leadership Visualization

<table>
<tr><th>Finland</th><th>China</th></tr>
<tr><td>Cluster: Nordic Europe

Also includes: Denmark and Sweden</td><td>Cluster: Confucian Asia

Also includes: Hong Kong, Japan, Singapore, South Korea, and Taiwan</td></tr>
<tr><td>Overall Profile

The Nordic European cluster has the lowest cluster scores for power distance, assertiveness, and in-group collectivism. Although countries in this cluster score low on in-group collectivism, they score high on institutional collectivism, indicating that, while they are generally self reliant, they promote group loyalty and encourage collective distribution of rewards in the workplace.

Societies in the Nordic Europe cluster want more reward and encouragement for performance excellence with more gender equality and equity. They want to be more group and family oriented with more pride, loyalty, and cohesiveness in their families, yet they desire less collective distribution of resources.</td><td>Overall Profile

The Confucian Asia cluster reflects high scores for power distance, institutional collectivism, and in-group collectivism. the high power distance score indicates that the societies in this cluster accept and endorse authority, power differentials, status privileges, and social inequality. Power is not expected to be distributed equally but is seen as providing social order and stability.

Societies in this cluster desire to decrease the level of power differentiation that exists. These societies do desire more reward and encouragement for performance, but prefer to be more future oriented. They also want their members to be kind, fair, friendly, and caring to each other. They also desire a lower level of male domination and gender role differences, but not as much as the average of other culture clusters.</td></tr>
<tr><td>Outstanding Leader Profile

• Performance oriented
• Visionary
• Very participative
• High integrity
• A degree of self-reliance
• Not status conscious

The Nordic European cluster scores higher than any of the other clusters on the participative leadership dimension, but lower than the other clusters in humane-oriented leadership dimension.</td><td>Outstanding Leader Profile

• Performance oriented
• Charismatic
• Not particularly participative
• Somewhat team oriented
• Face saving
• Status conscious

The Confucian Asia score of humane-oriented leadership is higher than most other clusters, as are the scores for autonomous and self-protective leadership. Although participative leadership is valued, the Confucian Asia cluster scores among the lowest of all culture clusters.</td></tr>
</table>

accepted cultural norms, as theorized by Hofstede, GLOBE, and others, that distinguish Chinese leadership tendencies from Finnish (Western) leadership tendencies. Chinese leaders are increasingly aware of the benefits of blending the best of Western and Eastern management practices. But Finland and China are still quite culturally distant, making it likely that the Finnish team in our example above would have to adjust their change management approach to better suit the cultural context and to get the best out of both worlds.

Do you think the Finnish team in our example would need to make the same adjustments if they were involved in a change initiative in Sweden instead of China? Finland and Sweden are not as culturally distant as Finland and China. GLOBE researchers place both Finland and Sweden in the Nordic European culture cluster, but there are differences between them. Hofstede ranks Sweden as being more feminine, more tolerant of uncertainty, more pragmatic (long-term orientation), and more indulgent than Finland (Figure 4.3), which may help explain differences in management styles between the two countries. "Management by Perkele" is an expression sometimes used to refer to a Finnish approach to leadership that favors swift decision making in contrast to the Swedish tendency toward prolonged pondering and consensus decision making (see Sidebar "Management by Perkele"). So even countries that are culturally close can have different tendencies when it comes to leadership and decision making.

It may not be entirely appropriate to generalize cultures based solely on the models such as Hofstede's dimensions and the GLOBE study. Many countries, such as the United States, China, India, Saudi Arabia, and Russia, exhibit characteristics linked to more than one culture. Nevertheless, and despite criticisms, Hofstede's and GLOBE's generalizations on cultures are widely applied across a variety of disciplines, including international business and management, to better understand how the world operates. Of course, you will want to use caution and careful consideration in generalizing national culture and how it influences leadership behavior and decision making to avoid negative stereotyping. And, again, in addition to national culture, regional, occupational, organizational, professional, and group culture can influence leadership behaviors and decision making. Not to forget individual differences such as age, gender, and personality types. Any of these might have important effects on leadership behaviors, decision making, and employees' expectations of leaders. Here we have focused on national culture characteristics, assuming that understanding them can improve your ability to match your change management strategy to the cultural context.

In the next chapter, we will look at how national culture can influence communication, as well as things you might need to consider when developing a culturally tuned change management communications plan.

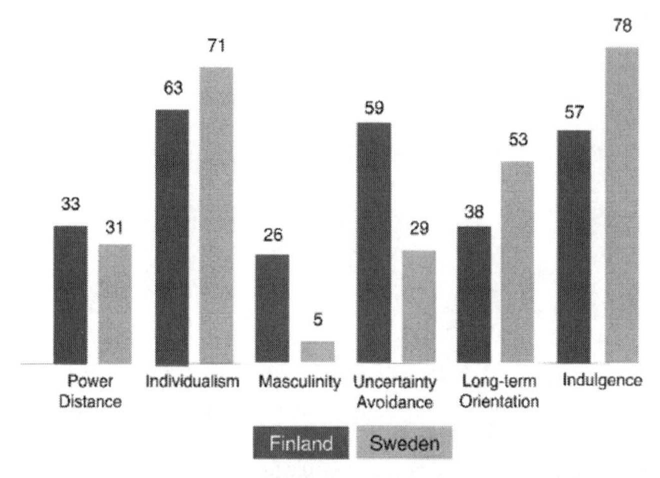

Figure 4.3 Cultural Profiles for Finland and Sweden Based on Hofstede's 6-Dimensions Model (Geert Hofstede, Gert Jan Hofstede, Michael Minkov, *Cultures and Organizations, Software of the Mind*, Third Revised Edition, McGrawHill 2010, ISBN 0-07-166418-1. © Geert Hofstede B.V. quoted with permission)

Management by Perkele

"Management by Perkele" is the Swedish stereotype of a Finnish approach to leadership that favors quick decision making, in contrast to the Swedish approach of consensus decision making involving the prolonged pondering of many alternatives in an effort reach a decision that is favorable to all parties. The term became famous in the 1980s, when Finnish companies began acquiring companies in Sweden.

As the story goes, during negotiations between some Swedish and Finnish directors, the Swedish approach became frustrating for the Finns. In his impatience with the Swedes—and in his hope to finally reach some kind of conclusion to the negotiations—the Finnish corporate director yelled, "Perkele," which is an old Finnish curse word.

There are a lot of negative connotations to "Management by Perkele." It is often considered a harsh, bullying, authoritarian style of management. Linus Torvalds, the founder of Linux, and a Finn, famously accused an Intel developer at Google of "Management by Perkele" for, as she said, turning the world of open source kernel development into a hostile work environment.

Even though the style is often connected to Finnish leaders, it can be recognized in leaders and managers in other countries as well. And of course, not all Finnish leaders manage by "perkele."

Key Points

- Just as organizational culture is intertwined with leadership, national cultural characteristics can also shape leadership perception and value judgments.
- A cultural analysis of leadership and decision making can enhance your cross-cultural capacity and improve your chances for collaboration in different cultural contexts.
- Cultural variables can influence things such as the speed of decision-making, acceptance of unpopular decisions, level of commitment for implementing a decision, and reactions to change in very different ways.
- Cultures are not monolithic, and, within each societal culture, people can range along the continuum of each cultural dimension, no matter the overall country ranking in the various cultural models (e.g., Hofstede, GLOBE, etc.).
- The perception of what constitutes outstanding leadership can vary from culture to culture.
- With awareness of how culture can influence leadership behaviors and employee expectations about leadership, the leader can start to build culturally mindful strategies to encourage appropriate interactions.
- In addition to national culture regional, occupational, organizational, professional, and group culture can influence leadership behaviors and decision making as well as individual differences such as age, gender, and personality types.

Want to Know More?

Visit Hofstede's website (www.geerthofstede.com) for a list of the cultural dimension scores of each country included in the research. The base culture data for six dimensions of culture (as presented in *Cultures and Organizations,* 3rd edition, 2010) Is downloadable in xls, doc, csv, and sav (SPSS) formats. Researchers can use them without asking for permission, but anyone considering commercial use should contact Geert Hofstede through the website.

Researchers wishing to work with Hofstede's data are strongly encouraged to read *Culture's Consequences,* 2nd edition (2001). Additional advice can be found in the research and Value Survey Model (VMS) section of the website.

You might also be interested in GLOBE publications such as *Strategic Leadership Across Cultures.* GLOBE publications represent results from a 20-year research program investigating the influence of national culture on organizational effectiveness, showcasing findings from studies in areas such as strategic leadership effectiveness across cultures. You can find a full listing of GLOBE books, articles, and presentations and proceedings on the GLOBE project website (www.globeproject.com).

Chapter 5

Getting It Right When Communicating Across Cultures

*The essence of cross-cultural communication has more to do with releasing responses than with sending messages. It is more important to release the right response than to send the right message.**

— Edward T. Hall

A Japanese businesswoman wants to tell her Canadian counterpart that she is not interested a particular deal. To be polite, she says, "This will be very difficult." The Canadian businessman, interpreting this to mean there are still some unresolved issues—not that the deal is dead—responds by asking how he can help resolve the issues. The Japanese businesswoman, believing she has sent the message that there will be no deal, is perplexed by the response. What went wrong?

Communication is fundamental to change management, not least because change is a collaborative activity. But communication does not necessarily result in understanding. Every communication has a message *sender* and *receiver,* and the received message is rarely identical to the sent message. The parties interpret each other's words and actions in terms of their own understandings, assuming that these are shared, when in fact they may not be. This is particularly true where intercultural interactions are concerned. Intercultural communication requires you

* http://www.azquotes.com/quote/747090

to be aware of cultural differences, because what may be considered acceptable and common in one culture might be offending or confusing in another country.

Communication styles vary enormously around the world, and the differences among them can become major sources of misunderstanding and frustration. Generally, the greater the difference in background (cultural distance) between *sender* and *receiver,* the greater the difference in the meanings attached to particular words and behaviors—and the greater the chance for misunderstanding and missed opportunities for cooperation.

There are two main schools in the study of communication: *process* and *semiotic* (Fiske, 2002). The process school of study views communication as the transmission of messages—that is, how senders and receivers code and decode messages, and how channels and media are used to convey accurate messages. The semiotic school sees communication as the production and exchange of meaning. It considers cultural differences between the sender and the receiver as a probable cause of communication failure.

"What is it that can be culturally relative in communication? The answer is, just about everything—all the aspects of what to say and how to say it" (Tannen, 1983). According to Tannen, there are seven levels of differences in communication across cultures (Table 5.1).

Unlike *intracultural* communication, in which senders and receivers generally understand the same ground rules around verbal and non-verbal communication, *intercultural* communication involves additional social and psychological variables that create uncertainty and ambiguity around the rules of engagement. Uncertainty can lead to miscommunication, so a key objective of a culturally tuned change management communication plan should be to reduce uncertainty. Unless uncertainty is reduced, it will be difficult for senders and receivers to communicate effectively. Of course, as with intracultural communication, some conflict and misunderstanding is inevitable. The goal is to mitigate misunderstanding even if we can't completely eliminate it.

5.1 Cultural Variability

One of the primary functions of culture is to define norms for interpersonal communication. Everyone's communication style (both verbal and non-verbal) is shaped by the values, norms, and thinking styles of the cultural group to which they belong. "Culture and communication are inseparable because culture not only dictates who talks with whom, about what, and how the communication proceeds, it also helps to determine how people encode messages, the meanings they have for messages, and the conditions and circumstances under which various messages may or may not be sent, noticed, or interpreted" (Samovar, Porter, and Jain, 1981).

Table 5.1 Levels of Communication Differences*

When to speak	When to speak is culturally relevant, and cultures differ in how silence is perceived and when it is appropriate. Those from cultures that expect more talk may perceive those who are more silent as uncooperative or disengaged. Conversely, those who talk less may perceive those who talk more as pushy and untrustworthy.
What to say	What is appropriate to say in different cultural contexts? Is it appropriate to ask questions? If so, what questions can be asked without causing offense? Telling stories is a universal way of communicating. But what can they be about, and when should they be told in different cultural contexts? A story that may be appropriate from the speaker's cultural perspective may not be so appropriate from the receiver's cultural perspective. When is it appropriate to use irony or sarcasm or to tell a joke?
Pacing and pausing	How fast should you speak? How long do you wait to speak after someone has spoken to be sure they have actually finished speaking and are not just pausing? How long should you wait between turns? Maybe the person who isn't speaking is just waiting for a long enough pause before they take their turn. But the person who is talking perceives a long pause as uncomfortable silence and continues talking. It all depends on the cultural context.
Listenership	What does good listenership look like? Does it mean holding a steady gaze, maintaining eye contact with the speaker?
Intonation	Intonation is made up of degrees and shifts in pitch, loudness, and rhythm of speech. There are cultural differences in how these little signals are used, and even small differences in intonation can throw an interaction completely off without the speakers knowing what caused the problem.
Formulacity	Every language is full of verbal expressions that are fixed in form, often non-literal in meaning, with attitudinal nuances. Formulaic language includes idioms, expletives, proverbs, and pause fillers (e.g., "Uhm," "Er") that non-native speakers might misinterpret or not understand. For example, "We hit it out of the ballpark" used in a business context would be understood by most Americans, but is likely to be meaningless to non-Americans (unless they live in a country where baseball is played or well understood).

* (Data drawn from Tannen, 1983. The Pragmatics of Cross-Cultural Communication. *Applied Linguistics*, 5[3]: 189–195)

(continues on next page)

Table 5.1 Levels of Communication Differences (cont.)

Indirectness	Communication in any culture is a matter of indirectness, to some extent. Only a part of the meaning is revealed in the words spoken; the largest part is communicated by hints, and the receiver must decode the message based on context, prior experience, and assumptions. People in high-context cultures are more aware than people in low-context cultures that much of what is meant cannot be said outright. This can create problems, even within a culture, of deciphering what is meant that is not said. Cross-culturally it can become a frustrating guessing game.

Cultural patterns of communication have generally developed over time based on geography, history, politics, and other influences. Individuals within a given culture are likely to misunderstand or ignore communications that do not conform to the expected societal pattern. The key to intercultural communication is understanding the other party well enough to make appropriate cultural adjustments to one's own communication style (see Sidebar "Why Native English Speakers Fail to Be Understood"). Foreign language skills are helpful, but simply being able to translate words or sentences from one language into another does not guarantee trouble-free communication. Being able to understand how language is used in different cultures and correctly interpreting the fundamental patterns of spoken and unspoken communication is at least as important.

5.2 Communication Styles

Good communication is a culturally bound concept. What is appropriate and effective in one culture may not be in another. And there is no guarantee that two people within the same culture will respond in exactly the same way; but broad-brush generalizations, if used mindfully, can give you some insight about what you are likely to encounter when communicating with people who belong to a culture different from your own. This insight can also help you develop a change management communication plan that fits the cultural context. Let's take a look at some of the communication styles.

5.2.1 Low-Context, High-Context

The concepts of "low-context" and "high-context" popularized by Edward Hall (1976) refer to how people communicate in different cultures based on

Why Native English Speakers Fail to Be Understood

English is one of the most widely spoken languages in the world today (see inset). But while native English speakers like to tell themselves that everyone else in the world now speaks their language, when travelling abroad, they frequently discover that their own English is often incomprehensible to business partners and colleagues (Hazel, 2016).

The perceived inability of native English speakers to refrain from using colloquialisms, slang, subtext, and cultural in-jokes has been found to result in resentment and suspicion. Foreign colleagues often resent the real or perceived lack of effort made by monoglot English speakers (Hazel, 2016).

If you are a native English speaker, reflecting on the difficulties others may have in understanding your style of English may be a good start toward becoming a more culturally mindful change agent and business partner.

15 Most Widely Spoken Languages (First and Second Language Speakers)

Rank	Language	1st Language	2nd Language	Total
1	Mandarin	960 million	193 million	1.15 billion
2	English	400 million	660 million	1.06 billion
3	Spanish	570 million	91 million	661 million
4	Hindustani	329 million	215 million	544 million
5	Arabic	290 million	132 million	422 million
6	Malay	77 million	204 million	281 million
7	Russian	153 million	113 million	267 million
8	Bengali	242 million	19 million	261 million
9	Portuguese	218 million	11 million	229 million
10	French	76 million	153 million	229 million
11	Hausa	85 million	65 million	150 million
12	Punjabi	148 million	?	148 million
13	German	76 million	52 million	129 million
14	Japanese	128 million	1 million	129 million
15	Persian	60 million	61 million	121 million

Source: Ethnologue (2017 20th edition)

the extent to which meaning is conveyed through explicit words or implied by context. As with the other cultural dimensions we have examined thus far, high–low context is a spectrum, and most cultures fall somewhere along the context continuum rather than completely at one end or the other.

Low-context communication is used primarily in more individualistic cultures, including Anglo countries and many European countries. Members of individualistic, low-context cultures have a tendency toward "over-explaining." Facts, descriptions, and preciseness are considered more important than context.

And, because everything needs to be spelled out explicitly, there is a higher reliance on written communication.

High-context communication is used primarily in more collectivistic cultures, which include many Asian, Latin American, Mediterranean, Slav, Arab, and West African countries. Members of collectivist, high-context cultures have a tendency toward "under-explaining." Context is more valuable than words, and messages are often implied. The listener is expected to be able to "read between the lines" to understand the unsaid. In fact, to explain everything and state meaning precisely can be interpreted as insulting. High-context communicators rely much more on non-verbal communication than low-context communicators. Body language, facial expressions, eye contact, and even silence are valued means of communication. Nonverbal behaviors can convey emotion and serve as emblems, speech illustrators, and conversation regulators. They can influence the receiver's understanding and acceptance of a spoken message. If the message receiver perceives a difference between the sender's verbal and non-verbal messages, the receiver is more likely to believe the non-verbal than the verbal communication.

It is also important to be aware that low-context languages tend to be writer-responsible, whereas high-context languages tend to be reader-responsible (Figure 5.1). These style differences can create cross-cultural misunderstandings in things like emails, business communications, job descriptions, and technical writing. And who is responsible for sending the message also varies by culture and can affect the expectations of both the writer and the receiver.

As a general rule, cultures with Western European roots (WEIRD societies) tend to rely more heavily on low-context communication. The rest of the world (non-WEIRD societies) is socialized more toward high-context communication. (The United States and Japan are perhaps the world's most extreme cases of low-context and high-context cultures, respectively.) Naturally, high-context

Writer-Responsible Language	Reader-Responsible Language
Languages in low-context cultures tend to be writer responsible. For example, English is a writer-responsible language. This means it is the responsibility of the writer to make sure the message is understood. Writing is clear, direct, and unambiguous. A good writer assumes no or little background knowledge on the part of the reader.	Languages in high-context cultures tend to be reader responsible. For example, Korean, Chinese, and Japanese are reader-responsible languages. This means the reader is responsible for deciphering the message, which is often not stated explicitly. This style can be confusing for a low-context reader who is expecting direct and explicit information.

Figure 5.1 Writer-Responsible and Reader-Responsible Language

communication can occur in a low-context culture—in a close-knit group, for example. And the use of low-context communication is becoming more commonplace in high-context societies due to globalization and Western influence. Understanding whether your intercultural interactions are with high-context or low-context communicators can help you make appropriate cultural adjustments to your own communication style and develop a communication plan that is more culturally tuned.

5.2.2 Direct–Indirect

The direct–indirect spectrum refers to the way speakers express their true intention in terms of needs, wants, and desires, and the differences between the two styles can provoke breakdowns in communication. For example, low-context (direct style) communicators can say "No" when something is not accepted, and to do so is generally not seen as impolite or offensive. By contrast, high-context (indirect style) communicators tend to avoid saying "No" in order to maintain a positive atmosphere. Think back to the example of the Japanese businesswoman saying "This will be very difficult" to communicate to the Canadian businessman that the hoped-for deal was dead. The message *sent* was not the message *received*. Each interpreted the same conversation differently, resulting in misunderstanding and confusion.

Individualistic (low-context) cultures tend to place greater emphasis on verbal clarity than collectivistic cultures (high-context). Direct and indirect communication styles illustrate the difference between high-context and low-context cultures. High-context (collectivistic) cultures have a preference for an indirect style of communication, in which verbal messages are designed to camouflage the speaker's true intentions, opinions, and needs. Indirect communication reflects a cautious attitude toward the expression of negative and confrontational messages. Low-context (individualistic) cultures generally prefer a more direct communication style, in which messages reveal the speaker's true intentions, opinions, and needs. Competent communicators are expected to say what they mean and mean what they say (refer to Figure 5.1).

5.2.3 Monochronic–Polychronic

The monochromic–polychronic spectrum refers to a culture's perspective about time, which, of course, has an impact on communication. Monochronic or sequential tendencies generally correlate to low-context, individualistic cultures. People in these societies think of time as a linear commodity, with activities placed along that line in a sequential order, in a logical, efficient way. By contrast, polychronic tendencies generally correlate to high-context, collectivistic cultures in which time is viewed as a flexible, constant flow to be experienced in the moment. Polychronic people are not slaves of a time line. They take a more relaxed approach to time, focusing more on the end result than on following planned time and schedules. Relationships are more important than tasks.

A culture's perspective about time can make a big difference in how its people relate to other cultures, and it can make planning a challenge. When you make an appointment with a person from a polychronic culture, it should be viewed as an "intention" to meet at that time. The actual meeting might be rescheduled or start earlier or later than planned, and it will take as long as needed to complete the goals for that meeting. But your polychronic colleague, business partner, or client might find it perfectly normal to take phone calls during the meeting or for others to interrupt it. If you have a monochronic orientation, you might interpret this type of behavior as rude or insulting. On the other hand, if you're polychronic, you might not understand why your monochronic colleague is so task driven and such a stickler for schedules and punctuality. Interactions between the two types can be problematic, and it is something you need to take into consideration when planning your change management activities.

5.2.4 Neutral–Affective

If you became angry or frustrated at work, would you say so? How would you express your feelings? Your answer will be very different depending on whether your culture is more affective or more neutral. The neutral–affective spectrum refers to how overtly a culture expresses emotion. Members of neutral cultures tend to keep their emotions carefully controlled and subdued, whereas members of affective cultures show their feelings freely and effusively. Japan, Poland, New Zealand, and Hong Kong are among the countries on the neutral end of the spectrum, whilst Kuwait, Spain, Saudi Arabia, Russia, and Italy are among the countries on the affective end of the spectrum.

Some manifestations of affective or neutral cultures fall under the umbrella of "cultural etiquette"—such as making direct eye contact, touching other people, the amount of space to keep between people, and assumptions about privacy. For example, Americans tend to feel most comfortable at arm's length in a social interaction, whereas Latin Americans would consider that distance unfriendly. Touching in public is commonplace in Latin American cultures, but not so much so in Asian cultures, where it is seen to be impolite. Knowing whether a culture is neutral or affective (emotional) can be especially helpful when you are preparing for intercultural interactions.

5.2.5 Informal–Formal

The informal–formal spectrum relates to power distance, which you will recall is the extent to which employees accept that superiors have more power than they do. Speaking can be formal or informal, depending on cultural norms. Informal cultures assume that everyone is equal, so people in these cultures speak to everyone in more or less the same way. In more formal cultures, it is assumed that there is a hierarchy among people, and they are expected to a follow certain protocols depending on to whom they are talking.

Formal cultures are generally collectivist, whereas social structures are centralized, with responsibility concentrated at the top of the structure. By contrast, informal cultures are generally individualistic, where the social structure is decentralized with responsibility going further down the structure rather than being concentrated at the top.

Before dealing with someone from a formal culture, you should make an effort to learn the required etiquette for the interaction, because maintaining the power distance is important.

Drawing on Hofstede's 6-Dimensions Model (Hofstede, 2011), Table 5.2 (beginning on page 74) summarizes how each of the cultural dimensions can influence communication styles and preferences, highlighting some of the things you should consider when developing a culturally tuned change management communications plan.

Naturally, other variables may affect communication, including task demands, receiver characteristics, organizational and occupational or professional culture, technology acceptance, and individual preferences. Individuals within a culture may not entirely fit into their culture.

5.3 Lewis Cultural Types Model

British linguist Richard Lewis's (2006) cultural types model is a good tool for gaining a better understanding of how to communicate with people from different cultures. It provides a behavioral comparison that plots countries/peoples in relation to three categories (see Figure 5.2):

- **Linear-actives.** Peoples who plan, schedule, organize, pursue action chains, do one thing at a time. This group includes the English-speaking world—North America, Britain, Australia, and New Zealand—and Northern Europe, including Scandinavia and Germanic countries.
- **Re-actives.** Peoples who prioritize courtesy and respect, listening quietly and calmly to their interlocutors and reacting carefully to the other side's

proposals. This group includes countries in Asia (except the Indian subcontinent) and Finns.

- **Multi-actives.** Peoples who do many things at once, planning their priorities not according to a time schedule, but according to the relative thrill or importance that each appointment brings with it. This group is more scattered and includes Southern Europe, Mediterranean countries, South America, sub-Saharan Africa, Arab and other cultures in the Middle East, India and Pakistan, and most Slavs.

While the three types are distinctive, each possesses behavioral elements from the other two categories. It is a question of which one is dominant.

In addition to insights gained from Hofstede, GLOBE, and Hall, the Lewis Model is useful for analyzing how different cultures are programmed, understanding your own personal orientation, and identifying the commonalities among different cultures so you can exploit them to improve your intercultural interactions and change management communications plans to suit different cultural contexts.

Now, let's return to our Finnish–Chinese example from the previous chapter. Let's assume the Finnish team has now assessed communication styles and preferences through the lens of the cultural dimensions. How close or distant are Finland and China in terms of communication styles? (See Table 5.3, page 77.)

This analysis gives the Finnish team insight into the cultural values that underpin Chinese communication. This enables them to anticipate that they will be working with people who are more accustomed to an indirect and nuanced communication style intended to preserve harmony and protect face.

Lewis Model of Culture		
Linear-Active	**Multi-Active**	**Reactive**
Results-oriented	Relationship-oriented	Harmony-oriented
Job-oriented	People-oriented	Very people-oriented
Cool	Warm	Courageous
Factual	Emotional	Accommodating
Decisive Planner	Impulsive	Compromiser
Written word important	Spoken word important	Face-to-face contact important
Restrained body language	Unrestrained body language	Subtle body language

Source: CrossCulture (www.crossculture.com)

Figure 5.2 Characteristics of Linear-Actives, Re-Actives, and Multi-Actives

Most Finnish workers are task oriented and very direct in their communication. So, how should the Finnish manager and her team adapt their style of communication when working with their Chinese counterparts?

Here are some of things the Finnish team might need to do:

- Be ready for some things to remain unclear.
- Be aware of body language. Gestures and body language may be important parts of the message.
- Give a straight "No" only when absolutely necessary.
- Emphasize the context.
- Invest time in forming or deepening relationships while you are preparing for the real work to start.
- In meetings, create the right atmosphere before getting down to business.
- Be aware that if a strong relationship doesn't exist, the Chinese may take a "wait-and-see" approach, requiring more visits, interaction, and especially time before real trust develops.

The Finnish team might wrestle with Chinese indirectness, and they may need to adjust by using a more indirect way of asking questions and accommodating the Chinese so they do not lose face. However, the possibility of being straightforward with the Chinese is also an option. Frankness is a typical Finnish value that may be accepted in certain situational contexts. It is possible to mutually learn that the same polite values and attitudes can be expressed in different ways.

Again, the example used here is fairly simplistic based on national characteristics, which are just a first step in understanding individuals (see Sidebar "Stereotypes and Generalizations," page 76). The point is to illustrate how the cultural research can be put into practice and to get you to think about how you can use it to develop more culturally sensitive communications plans and improve your intercultural interactions when managing change across cultures.

Table 5.2 Influence of Culture on Communications

Cultural Dimension	Things to Consider
Power Distance High power distance cultures tend to more formal communication. This can mean that managers often spend a lot of time monitoring routine messages. Huang et al. (2003) found that power distance can influence the choice of communications channels. For example, in high power distance cultures, email may not satisfy the requirements for symbols and cues showing status and respect. In low-power-distance cultures, however, email was more acceptable because the information was all that was required. The lack of symbols and cues was not considered a negative effect on its use. Members of low power distance cultures tend to be independent workers and are likely to have more input into decisions about communications content and the mechanism of delivery. Also, low power distance cultures tend to provide an environment that better supports multi-level distribution of data, information, and certain types of knowledge.	Based on your observations and experience with the organization, ask yourself: • Does communication tend to be more one-way (top-down) or two-way (top-down and bottom-up)? • Is there a preference for a more direct or more indirect style of communication? • Who should be involved in determining the content of communications and the mechanisms through which messages will be delivered? Should employees have input? What are the implications for the change management communications plan?
Individualism/Collectivism Communicators from individualistic societies are socialized to a low-context orientation. They tend to be more direct in their communication and to place less emphasis on the thoughts, feelings, and actions of others. Individualists emphasize task performance, whereas collectivists emphasize relationships. Communicators from collectivist cultures place more emphasis on high-context (indirect) communication and attribute meaning to both the context and the receiver's orientation. Message content is often embedded in the context of the communication, so the receiver needs contextual cues to interpret the message properly and continually looks for cues in communication (Hall, 1976). People from collectivist cultures are likely to	Based on your observations and experience with the organization, ask yourself: • Is culture high-context oriented or low-context oriented? How does your own orientation compare with the organization's orientation—high with high, high with low, or low with low? What are the implications? (If everyone has a similar orientation, then communication might be easier to some extent.) • Is there a preference for more direct (explicit) or more indirect (implicit) communication? How does your own style compare with the organization's tendencies—direct with direct, direct with indirect, or indirect with indirect? What are the implications? • Should communication reflect a more self-enhancing or a more self-effacing tone?

(continues on next page)

Table 5.2 Influence of Culture on Communications (cont.)

Cultural Dimension	Things to Consider
prefer synchronous communication, because it enables them to better understand the receiver's reactions to the message and make necessary adjustments. This suggests that collectivists emphasize more two-way communication, more personal communication, and more frequent com-munication, especially to coordinate activities and help clarify decision processes.	• Is there a high usage of non-verbal communication? What are the implications for written communication? • Which channels/mechanisms of communication are most effective for the cultural context? • Do people tend to be more task oriented or more relationship oriented? What are the implications for creating key messages and the mechanisms that should be used to transmit those messages?
Uncertainty Avoidance Communication is needed to reduce uncertainty and equivocality. Uncertainty avoidance drives people in lower uncertainty-avoidance societies to communicate in ways that are less rich than would be acceptable to members in higher uncertainty avoidance cultures. In higher uncertainty avoidance cultures, it is important to avoid lengthy delays in communicating, and messages should emphasize things such as job security. Receivers in cultures with a low level of uncertainty avoidance might be more accepting of information that is ambiguous or whose intent requires effort to understand. By contrast, receivers in high uncertainty avoidance cultures may be confused by or reject messages that are ambiguous or lack clear intent.	Based on your observations and experience with the organization, ask yourself: • Is there a tolerance for ambiguity in communications? What is the implication for key messages? • Is there a strong desire to explain everything as much as possible, or are people comfortable with less detail? • Is there a preference or need for more frequent or less frequent communication based on the degree of uncertainty avoidance? What is the implication for the timing of communications? • What type of communication is preferred? Do people tend to embrace or reject new communications channels/technologies?
Masculinity/Femininity In masculine cultures, there tends to be a greater emphasis on clarity in communication. In feminine cultures, there is more use of nonverbal com-munication. Feminine communication styles tend to be cooperative, and masculine styles tend to be competitive in tone.	Based on your observations and experience with the organization, ask yourself: • Does communication need to be more assertive or more collaborative in tone? • Should response styles reflect assertiveness or modesty?
Long-Term/Short-Term Orientation In more short-term oriented cultures, key communication should emphasize immediate rewards; in more long-term oriented	Based on your observations and experience with the organization, ask yourself: • Does the company emphasize short-term wins or long-term goals? What are the implications

(continues on next page)

Table 5.2 Influence of Culture on Communications (cont.)

Cultural Dimension	Things to Consider
culture, emphasize is usually on development and advancement opportunities.	for change management communications? • Are relationships more important than tasks, or vice versa?
Indulgence/Restraint Indulgent cultures place more importance on freedom of speech and personal control. In restrained cultures, there is a greater sense of helplessness about personal destiny. In the workplace, this is likely to have an impact on how willing or unwilling employees are to voice opinions and give feedback. More indulgent societies tend to have a more optimistic outlook on life, and this is reflected in their behavior and style of communication.	Based on your observations and experience with the organization, ask yourself: • Do leaders and employees tend to focus more on the positive aspects or the negative aspects of the change? What are the implications for key messages? • Are employees willing or unwilling to voice their opinions and provide feedback?

Stereotypes and Generalizations

Richard Lewis (2006) says that despite globalization people are still rooted in their national and regional backgrounds. But when you mention national characteristics someone will inevitably tell you they know someone from the culture you are discussing and they are nothing like what you describe. Then they accuse you of stereotyping.

Culture is complex, and describing national characteristics is just a first step in understanding the individual. As Lewis points out, anyone you deal with is a mixture of many different kinds of experience, and their national culture is just one of the variables that influence the way they think and act. Just as important are ethnicity, gender, generation, religion, social class, upbringing, education, and corporate experience. Generalizations based on national characteristics are simply a platform we can use to drill down to the individual, and this has proved to be enormously helpful in bringing international negotiations and projects to a successful conclusion.

Table 5.3 Comparison of Finland and China Based on Hofstede's Cultural Dimensions: Implications for Communication

Finland	China
• Low power distance • Individualistic • Uncertainty avoiding • Feminine • Short-term oriented (normative) • Indulgent **Communications Profile** As an individualistic and uncertainty-avoiding culture, Finns are usually very direct and explicit in their communication. They do not waste words or risk misunderstandings through diplomatic or roundabout (implicit) ways of saying things. Requests are delivered as orders. Finns are task oriented and monochromic. Punctuality is important. In meetings, they like to get down to business quickly, and they sometimes keep notes, transcripts, and records to help them focus on exactly what has been said and agreed upon. Finns live and work by rules. No contract will be signed without full consideration of the implementation and effects. This means that, while they are flexible during a constructive negotiation process, they resist changes to agreements and avoid breaking rules.	• High power distance • Collectivistic • Uncertainty accepting • Masculine • Long-term oriented (pragmatic) • Restrained **Communications Profile** As a high power distance and collectivistic cultural, being indirect helps the Chinese to protect and save face. The aim of all constructive communication is to avoid situations in which someone loses face within their group. The interpreters who are widely used in international negotiations are sometimes the scapegoat, taking the blame for "inaccurately translating" messages that may have caused embarrassment. Body language and gestures are commonly used and may be an important part of the message being delivered. Chinese are relationship oriented and tend to be polychromic, but lateness for meetings and appointments is unacceptable as it is seen as "stealing" time from your host. It is important to protect relationships, even if it risks missing deadlines. In meetings, it is important to create the right atmosphere before getting down to business.

Key Points

- Communication does not necessarily result in understanding. Every communication has a message *sender* and *receiver,* and the received message is rarely identical to the sent message.
- Intercultural communication requires you to be aware of cultural differences, because what may be considered acceptable and common in one culture is offending or confusing in another.
- Communication styles vary enormously around the world, and the differences between them can become major sources of misunderstanding.
- Unlike intra-cultural communication, where senders and receivers generally understand the same ground rules around verbal and non-verbal communication, intercultural communication involves additional social and psychological variables that create uncertainty and ambiguity around the rules of engagement.
- The key to intercultural communication is understanding the other party well enough to make appropriate cultural adjustments to your own communication style.
- Cultural characteristics influence communication style preference (e.g., direct or indirect, neutral or effective, formal or informal, etc.).
- High-context communication is favored in collectivistic cultures, low-context communication is favored in individualist cultures.

Want to Know More?

You can learn more about the Lewis Model from Cross Culture (www.crossculture.com). They also offer coaching and consultation on the use of the model.

To learn more about how different cultures express emotions in the workplace, you might be interested in reading *Riding the Waves of Culture* by interculturalist Fons Trompenaars (Trompenaars and Hampton-Turner, 2012).

Chapter 6

Influence of Culture on Group Identification, Performance, and Motivation

*The stranger sees only what he knows.**

— African proverb

Teams that straddle geographies and cultures have become part of working life. But cross-cultural working can be fraught with the unexpected, even when nations might superficially seem similar. Think back to the example of Finland and Sweden. Both countries fall into GLOBE's Nordic Europe culture cluster, and given their geographic proximity, you might expect culture to influence their ways of working in very similar ways. However, as we have seen, there are some marked differences between the two cultures when it comes to ways of working. Cultural differences between groups of people are not necessarily a problem, but when problems do occur, they can create difficulties in terms of teamwork, communication, motivation, or coordination. There is no single "best practice"

* https://medium.com/intercultural-mindset/28-quotes-that-will-level-up-your-intercultural-communication-skills-57790f649d97

for dealing with this, because how individuals interact in a group is highly contextual. Typically, leaders can choose a single management style when working in a single country. For example, in individualistic, low power distance countries such as Canada or Australia, using a participative leadership style can result in higher employee engagement and productivity. On the other hand, choosing to apply this style in countries that view power differently could be a blunder.

Power distance can also be a factor in team or work group responsiveness. High power distance culture employees are generally more willing to accept orders from supervisors without challenging or questioning them. This can reduce the amount of time that needs to be spent on "selling" the change and actions required in relation to the change, and it can potentially increase the speed of implementation. That is, responsiveness to the change and change-related actions increases with power distance (Kirsch et al., 2011).

6.1 Culture and Group Membership

Have you ever been in a situation in which a group of people treated you differently because they did not consider you part of their group? Have you ever wondered why this happens? We all belong to groups identified by traits such as age, gender, profession, ethnicity, and culture. And we tend to interact and favor our own group (in-group) more than groups to which we don't belong (out-groups). This is referred to as *in-group–out-group bias, in-group bias,* or *intergroup bias* and can be expressed in our evaluation of others, in allocation of resources, and in many other ways (Aronson et al., 2010). The norms, goals, and values of an in-group shape the behavior of its group members (Triandis, 1989). In-groups give people a sense of familiarity, trust, and personal security, because members of the group share traits such as language, culture, and a history of shared experiences (Hui, 1988). Virtually any basis for common categorizations can function as the basis for in-group favoritism.

We have seen that individualism–collectivism (Hofstede, 1980) is a cultural dimension that differentiates contexts in which individuals are socialized to be independent—driven by their own attitudes, beliefs, and convictions—or interdependent, relying on the dictates of the group in deciding on actions. The independent-self (individualism) is a more typical socialization pattern in Western Europe and North American societies, whereas the interdependent-self (collectivism) is a more typical pattern in much of Asia, Africa, Eastern and Southern Europe, and South America. Members of collectivistic cultures are more likely to define themselves in terms of their group memberships, whereas members of individualist cultures are more likely to define themselves in terms of their unique individual attributes (see Figure 6.1).

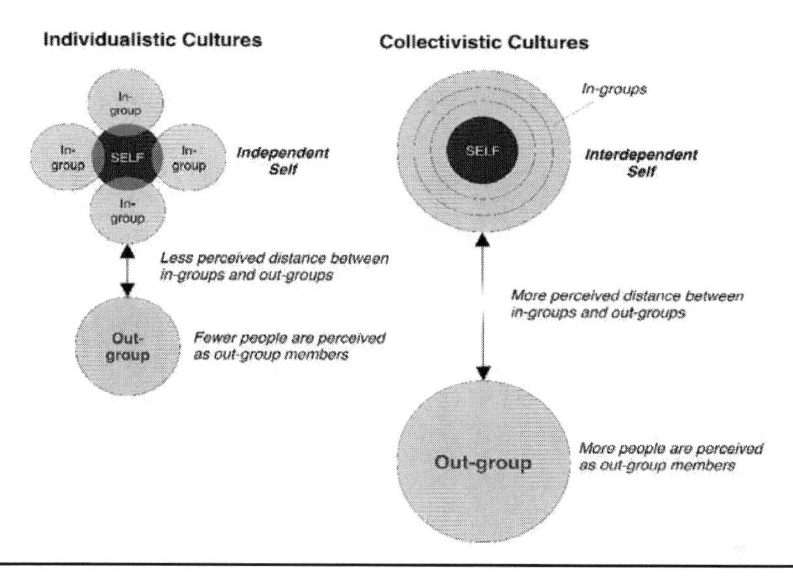

Figure 6.1 Differences Between Individualistic and Collectivistic Perceptions of Self and In-Groups and Out-Groups

Researchers have found a correlation between in-group/out-group distinction and the level of individualism or collectivism that exists in a culture (Bond, 1986; Bond and Hewstone, 1988; Hsu, 1988; Hui and Triandis, 1986; Triandis, 1986, 1995). More recently, researchers have hypothesized that there is also a middle group whose members are not perceived or treated as either in-group or out-group, but move into either group depending on the context of the situation and/or culture (Allred et al., 2007)—the implication being that people from collectivistic cultures do not necessarily perceive or treat people who are not in-group members as members of an out-group.

In addition, they proposed that the concept of a middle group also appears in individualistic cultures. For example, when Americans meet a person for the first time, they generally do not immediately categorized that person as in-group or out-group (unless they fall into a group against which the person already has a strong prejudice). They treat them in a less categorizing manner until they are able to form a more clearly defined attitude toward them. In comparing Americans and Chinese, the researchers (Allred et al., 2007) found that Americans tend to have a relatively large middle-group, whereas Chinese have larger in- and out-groups. The Chinese also have multiple layers of in-group members in comparison to Americans, and they tend to exhibit more extreme responses toward in-group and out-group members. However, even if the Chinese do feel more negatively toward out-group members, they may not state this directly, given the strong influence of the Confucian ideal to be kind and tolerant. And it is interesting that, according to the research, collectivistic cultures are less likely than individualistic cultures to label negative out-group behavior as a character flaw, possibly because they place less emphasis on individual attributes compared with individualistic cultures.

In-group bias has been aligned to a tendency to withhold praise and rewards from out-group members, and preference for the in-group can lead to greater tolerance of in-group behaviors that breach social codes, but swifter condemnation of any out-group behaviors that violate those same norms. When in-group–out-group bias occurs across cultures or ethnic groups, it can lead to ethnocentrism—that is, a belief in the superiority of one's own group.

Whenever we interact with other people, we face the problem of social uncertainty. Forming a mutually committed relationship (in-group) that is essentially closed to outsiders is often a response to problems of social uncertainty. Although in-group favoritism can provide advantages for the group, it can also be a liability. For example, when people strongly practice in-group favoritism, they are not able to fully exploit potential outside opportunities because they are too reluctant or unwilling to deal with outsiders. Conversely, in a society characterized by openness to many social relations with outsiders, there is more of an incentive to lower the barriers and act in a manner that is fairer than in-group favoritism.

Change management is centered, in part, on the interactions between managers and employees and between various groups impacted by the change. These interactions can be defined as part of either an in-group or an out-group (and maybe even a middle-group). With the out-group, the leader–follower and/or intergroup relationships may not be as smooth and as mutually beneficial as they might be with in-group relationships. In cross-cultural settings, it is worthwhile investigating the impact of intergroup dynamics on collaboration, knowledge exchange, and motivation. This is especially crucial where distributed teams and work groups are involved in the change initiative.

6.2 Group Member Performance

Team leaders and work group managers need to exercise influence effectively to manage change in organizations. This can be a tough proposition when you're dealing with people from different cultures, especially if those cultures are highly dissimilar to your own. Naturally, individuals and groups usually behave in the way most familiar to them. As the anonymous saying goes, "We don't see things as they are, we see them as we are." As this happens, the cross-cultural clashes and conflicts can begin to emerge. For instance, the reluctance of team members from individualistic cultures to prioritize the group's needs ahead of their own can result in trust issues (Costigan et al., 2007). Teams can fail when team members are unable to develop trust among themselves. And the influence behavior of managers reflects their own cultural values and traditions too. For example, compared to the Swiss, the Chinese value collectivism and power distance more than assertiveness. These differences in values suggest that Swiss managers may be more inclined to use direct confrontation to resolve problems and conflicts when they arise, which may backfire when working with people from high-context, collectivistic cultures who value harmony and protecting "face."

As we have seen, masculine cultures are based on values that favor a materialistic, aggressive, competitive, and achievement-oriented attitude. Feminine cultures value cooperation, collaboration, and human development and are generally more nurturing and caring. This means that team members from feminine cultures might expect or need a higher level of support from supervisors than do team members from more masculine cultures. Other aspects of group interactions may be impacted as well. For example, in collectivist societies it would be counterproductive and inappropriate to openly discuss certain issues, such as a person's performance. Where team members from more individualistic cultures might be accustomed to direct feedback on their performance, a more subtle, indirect way of communicating feedback might be more appropriate in collectivist cultures.

Participation and accountability are also strongly related to the ways in which people have been socialized to act based on the acceptable norms and standards of their culture. This influences their expectations relative to work roles and the supervisor–subordinate relationship. The concepts of employee participation, involvement, and industrial democracy originated in (WEIRD) low power distance cultures and are therefore compatible with values of those cultures, but not necessarily transferable to high power distance cultures. Although a power-sharing technique may work well in Switzerland or Sweden, it might find resistance in Mexico. However, where individualistic cultures focus more on individual performance, teamwork and collaborative change would probably be easier to introduce in collectivistic cultures, such as Mexico or China, because the team's objectives take precedence over individual goals and needs. And in collectivist cultures, the team members tend to have clearer ideas of performance objectives, better role clarity, and higher levels of accountability.

6.3 Group Member Motivation

Everyone understands the concept of motivation. We all know what it feels like to be motivated, and we all know what it feels like to be discouraged. But what we regard as motivation is closely related to our own culture, and often what is a motivating factor in our own culture can end up being de-motivating for people from another culture. The best motivational strategies vary depending on the cultural context. We can make some assumptions based on the cultural research about what may or may not motivate employees or groups from different cultural backgrounds, for example:

- Employees from individualistic cultures are more likely to be motivated by opportunities for greater autonomy and personal promotion, whereas employees from collectivistic cultures are more likely to be motivated by group incentives or receiving group support.
- Employees from low uncertainty avoidance cultures are likely to be motivated by opportunities for job role changes and promotions, whereas employees from high uncertainty avoidance cultures are more likely to be motivated by factors that ensure job security.
- Employees from cultures that are short-term oriented are more likely to be motivated by immediate rewards, whereas employees from cultures that are long-term oriented are more likely to be motivated by career path opportunities.

Of course, incentives and dis-incentives are often used as tactics for managing resistance to change. When you are managing change across cultures, you need be aware of what is important to employees and groups of different cultural backgrounds and then design appropriate motivation mechanisms.

Drawing on Hofstede's 6-Dimensions Model, Table 6.1 summarizes how each of the cultural dimensions can influence group and team dynamics, highlighting some of the things you should consider when developing a culturally tuned change manage communications plan.

Collectivism and power distance are correlated to the level of employee involvement and teamwork. Collectivism and uncertainty avoidance have an impact on accountability and role clarity. Team members in collectivistic cultures are more likely to have a clearer understanding of overall performance goals and how they fit into the bigger picture. In collectivistic cultures, team members have a clearer idea of performance objectives, better role clarity, and higher levels of accountability.

Cultures that are high in masculinity exhibit less expression of positive emotion (passion and drive). The nurturing, caring, and supportive environment of feminine cultures might stimulate more passion and drive in group members and employees because they feel more supported in their needs and endeavors. Uncertainty avoidance and masculinity are correlated with negative emotions (fear and anger). What are the implications for change management?

Table 6.1 Influence of Culture on Group and Team Dynamics

Cultural Dimension	Things to Consider
Power Distance The relationship between employees and their direct supervisor or team leader can be affected by the level of power distance and the degree of collectivism and masculinity within the culture. In high power distance cultures, team members may be unwilling to participate in decision making. They tend to be more oriented to tasks imposed by the manager and feel safe when they act following strict rules and procedures. As long as team members perceive a high power degree imposed by their manager, they will feel motivated to fulfill the tasks set. Group members from low power distance cultures generally want to be proactive, to be able to express their opinions, and to be able to ask for help without fear. They are more likely to be motivated by their involvement in decision making, and persistence of effort will generally be maintained as long as they feel involved in achieving the company´s goals, the manager does not exercise excessive control over them, and rewards are fair and consistent with the performances achieved.	Based on your observations and experience with the organization, ask yourself: • What is the power distance between managers and employees? • Do employees want and expect to be involved in decision making-related tasks, or are they more comfortable when the team manager makes task assignments? How close or distant are group member expectations from your own leadership style? • What are the implications for your change management plans when the change involves groups from both high and low power distance cultures? What approaches will you need to use to keep everyone engaged and motivated? How will you balance being more hands-off with low power distance group members against being more hands-on with high power distance group members?
Individualism/Collectivism Group members from collectivist cultures channel their effort to team goals, and that effort will generally be maintained as long as they feel they are a part of a group and feel they can rely on other members of the group. By contrast, group members from individualistic cultures are more oriented toward fulfillment of individual tasks due to their tendency toward self-sufficiency and self-motivation. Attention and effort are likely to be maintained as long as the tasks include new challenges, there are opportunities for advancement, and personal accomplishments are recognized and rewarded.	Based on your observations and experience with the organization, ask yourself: • Who are the in-groups, who are the out-groups, and who are the middle-groups? • How strong is the in-group–out-group bias? Are in-group members willing to work with out-group members? If not, what are the implications for your change management interventions? • How can you leverage in-groups from more individualistic cultures to facilitate engagement with out-group members? • How will you balance the motivation of the individual with motivation of the group? What are the implications for resistance to the change?

(continues on next page)

Table 6.1 Influence of Culture on Group and Team Dynamics (cont.)

Cultural Dimension	Things to Consider
Group members from individualistic cultures may find it easier to engage with out-group members than group members from collectivistic cultures do.	
Uncertainty Avoidance Members in low uncertainty avoidance cultures may be more open-minded about out-group members and find it easier to communicate with them than high uncertainty avoidance group members do. In low uncertainty avoidance cultures, group members are likely to be motivated by activities involving freedom of action, creativity, and diversity. They tend to focus on satisfying their desire for knowledge, discovery, and innovation. Motivation is more likely to be maintained when the team manager promotes a positive, informal work environment and involves group members in activities that arouse their curiosity and stimulate their creative ability. Group members in high uncertainty avoidance cultures are more likely to focus on maintaining certain situations, to rely on strict rules and group norms, or to want expert help. Motivation is more likely to be maintained when the environment is free of unpredictable situations and changes that might create tension, stress, and poor performance.	Based on your observations and experience with the organization, ask yourself: • What is the level of cooperation between in-group and out-group members? How can you leverage the open-mindedness of low uncertainty avoiding in-groups to facilitate cooperation with out-group members? • How will you strike a balance between the desire of low uncertainty avoiding group members for a more flexible environment and the desire of high uncertainty avoiding group members for a predictable environment with strict rules? What approaches will you use to keep both groups engaged and motivated without creating unnecessary frustration and conflict?
Masculinity/Femininity Group members from masculine cultures are geared more toward competitive activities involving substantial material rewards. Motivation and persistence in tasks is likely to be driven by the possibility of extrinsic rewards. Group members from feminine cultures are more likely to be geared toward flexible activities that allow a work–life balance. Motivation is likely to be maintained as long as the	Based on your observations and experience with the organization, ask yourself: • What potential problems do you anticipate may arise between group members from more masculine cultures and group members from more feminine cultures? • How will you balance the desire for competition and individual performance-based rewards with the desire for work–life balance and group wellbeing? What are the implications for

(continues on next page)

Table 6.1 Influence of Culture on Group and Team Dynamics (cont.)

Cultural Dimension	Things to Consider
team manager promotes an egalitarian environment and offers group members the opportunity to choose a work program customized to their needs.	resistance management and reinforcement of the change?
Long-Term/Short-Term Orientation In long-term oriented cultures, group members are more likely to be engaged in innovative activities with long-term impact, and motivated by personal and professional development based on a well-structured career plan. Long-tem orientation is also characterized by a high degree of tenacity. Members in short-term oriented cultures are more geared toward activities that do not involve major changes. Effort is focused more on actions that provide immediate results and rewards. They are less likely to be interested in sustaining long-term effort and attention.	Based on your observations and experience with the organization, ask yourself: • How will you balance the desire for quick results and immediate rewards with the desire for long-term career development? • How will you keep short-term oriented group members engaged and motivated on a long-term change initiative? What are the implications for resistance management?
Indulgence/Restraint Regarding the impact on work motivation, group members from indulgent cultures are more likely to be oriented toward tasks that take place in a positive and relaxed atmosphere, allow for freedom of opinion and expression, and promote networking between team members. Attention and effort is more likely to be maintained as long as time for leisure activities, relaxation, and socializing are respected. Group members in more restrained cultures are likely to be oriented more toward routine activities that do not involve a high degree of socialization and do not require expressing opinions. Their effort will be focused on actions that involve tangible rewards.	Based on your observations and experience with the organization, ask yourself: • How can you create an atmosphere that will support freedom of expression and promote networking without alienating those group members who prefer not to express opinions or be involved in a high degree of socialization? What are the implications for collaboration and trust building? What are the implications for individual and group motivation?

From this analysis, we can imagine that teams in collectivistic cultures will have significantly better team processes and resources. In highly individualistic cultures, the reluctance of team members to prioritize the group ahead of their own needs is likely to result in trust issues between group members and the team leader or supervisor. We would also expect team members to feel more supported, recognized, and rewarded in more feminine cultures. The relationship between team members and the team leader is likely to be impacted by the level of power distance and the degree of collectivism and masculinity.

Returning to our Finnish–Chinese example, let's assume the Finnish team has now assessed group identification, performance, and motivation through the lens of the cultural dimensions. How close or distant are the Finnish and Chinese cultures in terms of groups and group dynamics? (see Table 6.2).

This analysis gives the Finnish team insight into the cultural values that underpin Chinese group identification, group dynamics, and worker performance and motivation, enabling them to anticipate that the working style of Chinese groups and teams will be different from how groups and teams function in Finnish companies. How should the Finnish manager and her team adapt their expectations and approach when it comes to working with Chinese groups and teams?

Here are some of things the Finnish team might need to do:

- Know the key networks (in-groups/out-groups) inside and outside of work
- Operate through networks and personal connections whenever possible
- Give extra attention to maintaining relationships within the group
- Do things individually only when it's easier than doing them together
- Be ready to expect less recognition for individual achievements, and allow individual work to be credited as a team achievement
- Focus more on team objectives, and fulfill team tasks before working on individual tasks
- Aim for approval from the group when trying new approaches

Cultural factors can make or break a change project, so you will need to actively develop ways to cope with the cultural complexity, including differences in individual and group behavior, when managing change across cultures. Of course, members of other cultures involved may also need to make adjustments to their own ways of working to ensure productive and collaborative interactions between in-groups and out-groups.

Again, it is important to remember that the cultural dimensions are broad-brush norms, and other variables may affect group dynamics, performance, and motivation, including task demands, organizational and occupational or professional culture, technology acceptance, and individual preferences. And remember, individuals within a culture may not entirely fit into their culture, so it is important to take all contextual variables into account when dealing with cross-cultural groups.

Table 6.2 Comparison of Finland and China Based on Hofstede's Cultural Dimensions: Implications for Communication

Finland	China
• Low power distance • Individualistic • Uncertainty avoiding • Feminine • Short-term oriented (normative) • Indulgent	• High power distance • Collectivistic • Uncertainty accepting • Masculine • Long-term oriented (pragmatic) • Restrained
Group Identification, Performance, and Motivation Profile The Finnish working style is more individualistic. Finns aim for self-reliance and usually prefer to work alone, although they do accept collective action. Many Finnish companies have adopted team working, but Finnish teams tend to meet just often enough to ensure good communication and openness. Finns are more likely to be motivated by an environment of free expression and opportunities for rewards based on individual performance. Finns prefer to keep their personal life separate from their work life, and socializing among co-workers is limited. Finnish in-group members are open minded and fair when it comes to dealing with out-group members. If you are different, Finns will tolerate you and accept your difference, but you will never be quite accepted as the in-group as long as you do things differently.	**Group Identification, Performance, and Motivation Profile** The Chinese working style is more collectivistic. Chinese are socialized to be group oriented, and membership in workplace in-groups is very important. An individual's freedom of action, initiative, and social mobility may be greatly restricted by in-group obligations. Conformity is considered the norm, so one of the key motivators for the Chinese is approval from the group. Individual rights and identity are typically of less importance than keeping face and positive relationships within the in-group. Personal relationships within in-groups are the route to individual advancement, and in-group members depend upon it. Chinese in-group members are less open-minded about out-group members and are more likely to exhibit strong in-group favoritism.

Key Points

- Group identification, dynamics, performance, and motivation are influenced by national culture.
- Cultural differences between groups of people are not necessarily a problem, but when problems do occur, they can create difficulties in terms of teamwork, communication, motivation, or coordination.
- We tend to interact with and favor our own group (in-group) more than groups to which we don't belong (out-groups), which can lead to in-group–out-group bias.
- Preference for the in-group can lead to greater tolerance of in-group behaviors that breach social codes, but swifter condemnation of any out-group behaviors that violate those same norms.
- Members of collectivistic cultures are more likely to define themselves in terms of their group memberships, whereas members of individualistic cultures are more likely to define themselves in terms of their unique individual attributes.
- The concepts of employee participation and involvement originated in (WEIRD) low power distance cultures and are therefore compatible with values of those cultures, but not necessarily transferable to high power distance cultures.
- Often what is a motivating factor in our own culture can end up being de-motivating for people from another culture, so motivational strategies need to vary depending on the cultural context.

Want to Know More?

You can learn more about cross-cultural groups in Erin Meyer's insightful book *The Culture Map: Breaking Through the Invisible Boundaries of Global Business* (Meyer, 2014).

Chapter 7

Influence of Cultural Dimensions on Learning and Knowledge Transmission

*It is good to rub and polish our brain against that of others.**

— Michel Eyquen de Montaigne

Most of us unconsciously assume that our own ways of thinking and behaving are representative of how everyone else thinks and behaves, and this is also true when it comes to learning. Not everyone learns in the same way. Some of our learning behaviors are based on personal preference, but many of our learning behaviors are influenced by our culture. The unique challenge for change managers working across cultures is to understand which learning behaviors are based on cultural values—which maybe shouldn't be challenged—and which are more superficial and can be challenged for the sake of instructional design.

Of course, instructional strategies are also culture based, and an approach that works well in one culture many not necessarily work well or even be appropriate in another. For instruction to do the most good for learners, instructional

* BrainyQuote.com, Xplore Inc, 2018. https://www.brainyquote.com/quotes/michel_de_montaigne_102641, accessed June 12, 2018.

providers must be aware of the cultural backgrounds of the learners and how those cultures manifest themselves in learning preferences (Nisbett and Masuda, 2003). Knowing which instructional activities will be most effective for a particular group of learners in the cultural context and adapting the instructional strategies appropriately is crucial. And this means that instructors need to recognize and accept that their default approach may not be appropriate in every cultural context. Instructors must also become more aware of the cultural biases embedded in their own instructional designs, including the selection of learning activities, their presentation style, and their expectations of students. Ignorance of these biases could prevent them from seeing opportunities for more effective interaction with and a better experience for learners.

"Culture is so much a part of the construction of knowledge that it must underpin not only the analysis phase but all phases of the design process" (Thomas, Mitchell, and Joseph, 2002). Unfortunately, culture is often overlooked on change initiatives, because the analysis phase of instructional design is one of the most commonly skipped phases. But it is essential that instructional providers familiarize themselves with the learners' cultures throughout the planning and execution of learning activities, and even during the post-learning evaluation stage. Where appropriate, this might include involving a cultural expert as part of the learning design team (Thomas, Mitchell, and Joseph, 2002; Young, 2008).

The cultural dimensions are useful for understanding the spectrum of cultural differences that impact knowledge sharing and learning. Where individuals fall along these dimensions impacts both instructors and learners. Of course, it is important to remember that although cultural differences can be usefully described along the cultural dimensions, within any culture individuals will differ in how strongly they display the tendencies associated with each dimension.

Drawing on Hofstede's 6-Dimensions Model, Table 7.1 summarizes how each of the cultural dimensions can influence learning and knowledge transmission in different cultural contexts, highlighting some of the things you should consider when developing a culturally tuned training plan.

In single cross-cultural situations, in which the instructor is dealing with a culturally homogeneous group of learners, but from a culture different from the instructor's culture, the instructional design should be adapted as much as possible based on the cultural analysis, without compromising the integrity of the content and underlying instructional principles (Castro, Barrera, Jr., and Martinez, Jr., 2004; Rogers, Graham, and Mayes, 2007). For example, based on the cultural context, should the learning be predominately discussion based, or should it be more lecture/presentation based? Should the instructor expect students to express opinions or ask challenging questions, or are the students expected to accept the instructor's point of view? Should there be some flexibility in the course schedule, or should a strict schedule be enforced?

Table 7.1 Influence of Culture on Learning and Knowledge Transmission

Cultural Dimension	Things to Consider
Power Distance In low power distance cultures, there is a greater focus on student-centered instruction. Instructors are treated as equals to be engaged and even challenged, dialogue and discussion are critical learning activities, and students take responsibility for learning activities. There is often more of a preference for younger teachers. In high power distance cultures, there is a greater focus on teacher-centered instruction. Instructors are the primary communicators and are treated as unchallenged authorities. Teachers are often solely responsible for what happens in the instruction. There is often a preference for older teachers.	Based on your observations and experience with the organization, ask yourself: • How do instructors/trainers expect learners to behave? Do they expect them to share opinions and ask challenging questions, or do they discourage debate and discussion? • Do learners expect instructors to take a more authoritarian or a more participative approach? Do they expect the instructors to be experts who give them the information they need, or do they expect the instructors to also function as facilitators for student discussions and debate? • Is there a preference for highly structured or more flexible learning interventions? • Do students take initiative for their learning, or do they prefer the instructor to take responsibility?
Individualism/Collectivism In more individualistic cultures, instructors expect students to speak up, and the student's point of view is seen as a valuable component of learning. Learning how to learn is primary, and individual gain is the motivation for hard work. In more collectivistic cultures, students rarely speak up, and students are expected to accommodate the instructor's point of view. Learning how to do is primary, and the greater good is the motivation for hard work. When it comes to knowledge sharing, in more collectivistic cultures, in-groups are more likely to share what they know with their in-group members. Individualists, who do not have strong affiliations with in-groups, may not be willing to share knowledge even with their immediate work collectives unless it is in their personal self-interest to do so.	Based on your observations and experience with the organization, ask yourself: • Do instructors expect learners to be as independent as possible, only giving them essential guidelines and information to get them started on their learning assignments? Are learners expected to work alone or to collaborate with other students? • Do learners prefer to work alone or to collaborate with other students? Is peer support evident in the learning environment? What are the implications for the design of learning activities? • Do learners have an attitude of, "Show me and I'll learn," or "Tell me and I'll learn," or "Give me the instructions to read, and I'll learn"? What are the implications for learning formats (i.e., classroom or virtual, instructor-led or on-demand, etc.). • Is there a preference for face-to-face (high-context) knowledge sharing as opposed to online communities (low context)?
Uncertainty Avoidance In high uncertainty avoidance cultures, learning activities tend to be highly structured.	Based on your observations and experience with the organization, ask yourself: • Do learners expect clear, unambiguous

(continues on next page)

Table 7.1 Influence of Culture on Learning and Knowledge Transmission (cont.)

Cultural Dimension	Things to Consider
Teachers are expected to have all the answers, and ambiguity is to be avoided. The learning environment tends to be more stressful than in low uncertainty avoidance cultures. In low uncertainty avoidance cultures, learning activities are more open ended and include discussions and projects. Teachers can say, "I don't know," and ambiguity is considered a natural condition. The learning environment tends to be less stressful than in high uncertainty avoidance cultures.	instructions, or do they prefer the flexibility to be creative? Do the students require and expect a lot of guidance? • Is the instructor expected to have all of the answers, or is it acceptable for the instructor to say, "I don't know"? What are the implications for instructor selection (style, credentials, etc.)? • Are learners willing to participate in active experimentation, or do they prefer reflective observation (gather information and reflect from a "safe" distance)? What are the implications for the instructional design?
Masculinity/Femininity In more feminine cultures, students are praised, and collaboration is cultivated. Average is used as the norm, and failure is viewed as a growth opportunity. In more masculine cultures, only excellence is praised, and competition is cultivated. The best student is used as the norm, and failure is highly frowned upon.	Based on your observations and experience with the organization, ask yourself: • Is more emphasis placed on surface learning or deep learning? • Are learners motivated more by collaborative activities or competitive activities?
Long-Term/Short-Term Orientation In short-term oriented cultures, the benefits of deep learning may not be apparent or appreciated. Learners may place more emphasis on scores/grades and class rank (immediate goals), whereas in long-term oriented cultures, learners are more likely to place greater value on gaining a deep understanding of the subject matter.	Based on your observations and experience with the organization, ask yourself: • Is there a greater focus on surface learning to achieve immediate learning objectives or deep learning that will be more beneficial in the long run?
Indulgence/Restraint In more indulgent cultures, learners are likely to be more talkative and expressive and may ignore plans. Instructional activities are allowed to continue as long as they are useful. In more restrained cultures, learners are more likely to work quietly toward planned ends, and they tend to prefer procedures. Instructional activities start and stop promptly.	Based on your observations and experience with the organization, ask yourself: • Do learners expect the learning activities to be structured, adhering to a rigid schedule? Or do they prefer a more flexible approach? • Do learners prefer to work quietly, or do they prefer a lot of discussion with their fellow learners?

In multicultural situations, accommodating the various cultural backgrounds might be achieved by offering students alternative choices in learning activities and instructional formats (Irvine and York, 1995; McLoughlin, 2001), depending on the degree to which ignoring any deeply rooted cultural differences will affect learning. This requires an analysis of the cultural dimensions: Which ones are most important to consider? Which ones present the most difficulty for instructional adaptation? Which culturally based learning differences are more easily accommodated? Do some extremes in cultural backgrounds represent incompatible approaches for learning? Of course, the cost of the alternatives is also a consideration.

Returning once again to our Finnish–Chinese example, let's assume that the Finnish team has now assessed instructional styles and learning preferences through the lens of the cultural dimensions. How close or distant are the Finnish and Chinese cultures in terms of knowledge sharing and learning? (See Table 7.2.)

Table 7.2 Comparison of Finland and China Based on Hofstede's Cultural Dimensions: Implications for Communication

Finland	China
• Low power distance • Individualistic • Uncertainty avoiding • Feminine • Short-term oriented (normative) • Indulgent	• High power distance • Collectivistic • Uncertainty accepting • Masculine • Long-term oriented (pragmatic) • Restrained
Learning and Knowledge Transmission Profile	**Learning and Knowledge Transmission Profile**
Finnish learners more readily accept that learning can take place through their own discovery and construction, and the instructor is viewed and accepted as a facilitator to bring about that discovery by providing direction and guidance to help the learners map out their own learning path. Finnish learners tend to be comfortable with sharing opinions and posing challenging questions to the instructor, and the degree of participation and involvement during discussions confirms this preference.	Chinese learners view the instructor as an expert who bestows wisdom and shares experience for the benefit of the learners, and learners tend to receive that wisdom without questioning. Learners expect the instructor to provide an explanation of the learning points at the outset. Questions are likely to be directed to peers in private discussion rather than to the instructor. Chinese learners are usually uncomfortable with searching for their own answers through discussion.
Learners tend to be self-directed and are comfortable with being involved in mapping out the direction of their learning.	Learners prefer to have the instructor make the decisions about the direction of their learning.
Learners are likely to accept learning based on theories, constructs, and abstract ideas.	Learners prefer learning based on concrete facts, procedures, and precedents.

This analysis gives the Finnish manager insight into the cultural values that underpin Chinese knowledge exchange and learning preferences and tendencies. This enables the Finnish team to anticipate that learning activities may need to be adapted to accommodate Chinese learners. How should the Finnish manager and team adapt their expectations and approach when it comes to working with Chinese groups and teams?

Here are some of things the Finnish team might need to do in conjunction with Learning specialists:

- Develop a culture-based instructional strategy.
- Design learning from a teacher-centered, reflective observation perspective.
- When designing learning activities, pay attention to the fact that learners may grasp concrete learning a lot more quickly than abstract learning.
- Assign an "expert" to prescribe the structure and syllabus for each learning activity.
- Use face-to-face formats more and electronic formats less for learning and knowledge-sharing activities.
- Select instructors who are already comfortable with or can easily adapt to a teacher-centered strategy.

What else can the Finnish team do to accommodate Chinese learning and knowledge-transmission preferences? Which cultural characteristics are the most important to accommodate? What other things should they consider if they need to develop a learning- and knowledge-transmission strategy for a multicultural group of leaners, as opposed to just Chinese learners?

The ability to accommodate culturally based learning differences is becoming increasingly more important in this time of rapid globalization. A recent survey conducted by the American Society for Training and Development (ASTD, 2012) found that one of the reasons why 72 percent of multinational companies were dissatisfied with their global training initiatives was that these initiatives failed to take cultural differences in learning into account. This means that, as a change management professional, you must become more knowledgeable about the cultural differences and the intended and unintended consequences of instructional designs in order to develop learning strategies with greater cultural sensitivity. Researching the multicultural education and training challenges may lead to greater wisdom.

Key Points

- Some learning behaviors are based on personal preference, but many are influenced by culture.
- Instructional strategies are also culture based, and an approach that works well in one culture may not work well or even be appropriate in another.
- Culture is often overlooked because the analysis phase of instructional design is one of the most commonly skipped phases.
- Knowing which instructional activities will be most effective for a particular group of learners in the cultural context, and then adapting the instructional strategies appropriately, is crucial.
- The cultural dimensions are useful for understanding the spectrum of cultural differences that impact knowledge sharing and learning.
- Instructional design should be adapted as much as possible based on the cultural analysis but without compromising the integrity of the content and underlying instructional principles.
- Change management professionals must become more knowledgeable about cultural differences and the intended and unintended consequences of instructional design in order to develop more culturally sensitive learning strategies.

Want to Know More?

You might be interested in reading Richard Nisbett's landmark book, *The Geography of Thought: How Asians and Westerns Think Differently* (Nisbett, 2004).

Chapter 8

Understanding Resistance to Change Across Cultures

People don't resist change. They resist being changed.[*]

— Peter Senge

Organizational change is viewed as an individual-level phenomenon because it occurs only when the majority of people in the organization change their behaviors or attitudes (Whelan-Berry, Gordon, and Hinings, 2003). But people can be the biggest obstacle to change. Resistance is a natural reaction to change, and it can manifest itself in many ways. But while there are predictable, universal sources of resistance, our perception of and reaction to change can be influenced by our culture. For example, although resistance to change can be found in all organizations in all countries, resistance will be significantly greater in high uncertainty avoidance cultures. People from cultures that strongly avoid uncertainty look for anchors that ensure a desired level of certainty (Hofstede, 2005), and they are more likely to resist or reject change because it creates anxiety and threatens certainty. On the other hand, people from cultures that are more tolerant of ambiguity and uncertainty are less likely to strongly resist or reject change.

[*] https://www.goodreads.com/quotes/125720-people-don-t-resist-change-they-resist-being-changed

Of course, this doesn't mean that every person from a high uncertainty avoidance culture will always resist change or that every person from a low uncertainty avoidance culture will always embrace change. The response to change is influenced by many variables, including national culture, the organizational environment, and the individual's personality.

And it's not just the degree of uncertainty avoidance that influences reactions to change. According to Harzing and Hofstede (1996), cultures that are characterized by a combination of high power distance, collectivism, and high uncertainty avoidance—most Latin American countries, Korea, France, Greece, and Arab countries, for example—are more likely to encounter the strongest resistance to change. Countries low on power distance, high on individualism, and low on uncertainty avoidance—Anglo countries, Nordic countries, and Singapore, for example—are more likely to have lower levels of resistance to change.

One of the most common reasons people resist organizational change, regardless of their culture, is because they believe the change will lead to the loss of something valuable. Classical management theories have offered many suggestions on how to reduce resistance to change, such as raising awareness of why the change is needed and the risks of not changing, communicating more frequently, inviting employees to participate, developing stronger working relationships, and providing people with needed resources (Kotter, 1995; Kotter and Schlesinger, 1979; Kouzes and Posner, 1993).

There is nothing wrong with these approaches. But no approach is context free, and change practitioners must understand the advantages and disadvantages of these approaches and how to apply them in different cultural contexts. Analyzing the cultural context aims at making sense of employees' reactions to change initiatives and getting a better understanding of the intricate relation between distinct factors of resistance to change.

Resistance can range from apathy to aggressive resistance (Figure 8.1). Although every change is unique, there are cultural tendencies related to perceptions of and reactions to change.

Drawing on Hofstede's 6-Dimensions Model, Table 8.1 summarizes how each of the cultural dimensions can react to change and highlights some of the things you should consider when developing a culturally tuned change manage communications plan.

Taking the cultural dimensions together, we can see that the strength of resistance to change is likely to vary depending on the combination of cultural variables. The degree of resistance may also depend on the type of change that is taking place in cultural context. For example, restrained and uncertainty-avoiding cultures may resist technology changes more strongly than other types of change. But it is important to remember that these cultural dimensions represent a spectrum, and countries fall somewhere along the spectrum for each

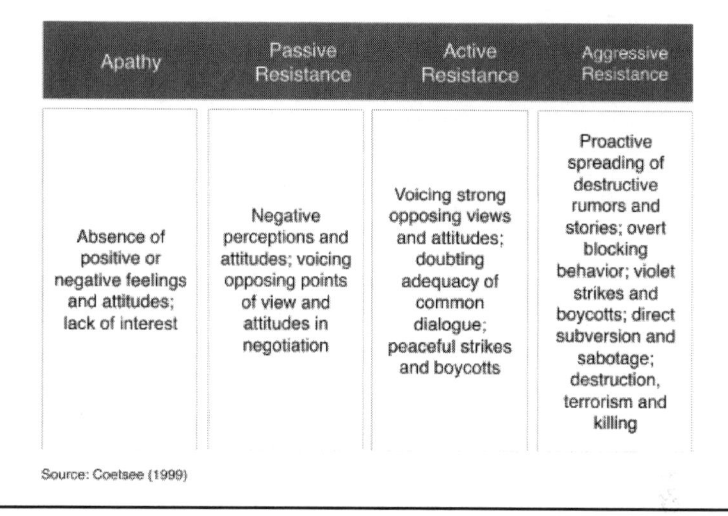

Figure 8.1 Forms of Resistance to Change (*Source:* Coetsee, L. 1999. *Resistance to Commitment.* Southern Public Administration Education Foundation. Reprinted with permission)

dimension rather than completely at one end or another. Therefore, it is important to take all of the contextual variables into account when anticipating and managing resistance to change. The cultural dimensions are only one way of understanding a very complex world.

Returning to our Finnish–Chinese example, let's assume the Finnish team has now assessed resistance to change through the lens of the cultural dimensions. How close or distant are the Finnish and Chinese cultures in terms of reactions to change? (See Table 8.2, page 106.)

This analysis gives the Finnish team insight into the cultural values that underpin Chinese employees' potential reactions to change. This enables them to anticipate that Chinese employees might be less resistant to change, or resistant in different ways than Finns. How should the Finnish team members adapt their expectations and approach with regard to resistance management in the Chinese organization?

Here are some of things the Finnish team might need to do:

- Give some recognition to the effects of outside events (fate or luck).
- Give people a method for how they can make success of the change feel inevitable.
- Recognize not only individuals' sensitivity to the change, but also that group pressure may cause them to resist the change even if they personally feel neutral to it or supportive of it.
- Show recognition for people's ability to understand the big picture.

- Build flexibility into the change management plans, and be prepared for people to return to previously made decisions based on new circumstances.
- Don't focus on narrow (short-term) results, but on how the change improves the "greater good."
- Be aware that Chinese managers may use implicit or explicit coercion to manage resistance to change, especially when speed is essential.
- Be ready for no one in particular to get credit for successes or failures.

Resistance management strategies require the skillful application of a number of strategies, often in different combinations, and choosing the right strategies is context dependent. You must give consideration to all of the contextual variables in play, including national culture, in order to develop resistance management strategies that are culturally sensitive and best suited to the situation.

Table 8.1 Influence of Culture on Resistance to Change

Cultural Dimension	Things to Consider
Power Distance In high power distance cultures, top management initiates the change, and subordinates are expected to implement it. Given the strict hierarchal structure of organizations in collectivistic cultures, less time needs to be spent on "selling" the change, and the change is less likely to be strongly resisted or rejected. But when resistance does occur, leaders who hold a lot of power and discretion may us coercion or manipulation to deal with the resistance. In lower power distance organizations, which are typically found in individualistic cultures, more bottom-up and inclusive processes are involved in initiating and managing change. As a result, there is a greater likelihood of professional disagreement, which could lead to strong resistance or rejection of the change. When resistance does arise, leaders tend to rely more on negotiation, partici-pation, and involvement as risk mitigation tactics and less on coercion and manipulation.	Based on your observations and experience with the organization, ask yourself: • Is the approach to change more top-down or more bottom-up? • Do employees generally accept mandates from change initiators unquestioningly, or do they expect a lot of information about why the change is necessary? What are the implications for the amount of time that needs to be spent on making the case for change? • What are the anticipated sources of resistance? • Which approaches are typically used to manage resistance to change—coercion and manipulation or participation and involvement? What are the benefits and drawbacks of the approaches?
Individualism/Collectivism In collectivistic cultures, resistance is more likely to be passive, and change is less likely to be resisted when it is perceived as being for the greater good. By contrast, in individualistic cultures, change is likely to be resisted if it does not serve the self-interests of individuals, and resistance is likely to be both passive and active. In collectivistic cultures, personal opinions are submerged by group opinions. Group pressure, then, is an important source of resistance. When change violates the benefits of most of the people in a group, resistance can occur. If the group resists change, individual members of the group who hold indifferent or positive attitudes toward the change will feel pressure and will choose to follow the group.	Based on your observations and experience with the organization, ask yourself: • What are the anticipated sources of resistance—groups or professional disagreements? What are the implications for resistance management? • Are individuals able to express resistance or do they follow the crowd even if they have a difference of opinion? Does this need to be addressed? If so, how can it be done in a culturally sensitive way? • Is resistance most likely to be apathetic, passive, active, or aggressive?

(continues on next page)

Table 8.1 Influence of Culture on Resistance to Change (cont.)

Cultural Dimension	Things to Consider
In individualistic cultures, professional disagreement from employees is a more frequent source of resistance. Resistance to change is more subject to individual reaction, rather than group reaction.	
Uncertainty Avoidance High uncertainty avoidance cultures are more likely to strongly resist or reject change, because change represents a threat to certainty. In fact, some uncertainty avoiding cultures may view with suspicion and resist the very notion of organizational change management. As a result, there is a lower tolerance for innovation, and the pace of change is likely to be slower. Any change that threatens job security is likely to be strongly resisted or rejected. In low uncertainty avoidance cultures, change is less likely to be strongly resisted, and there is likely to be a greater tolerance for innovation and a faster pace of change.	Based on your observations and experience with the organization, ask yourself: • What is the general reaction to change? Is it seen more as a threat or an opportunity? • What is the appetite for change? Should the pace of change be faster or slower? • Will the change impact job security? If so, how strongly will employees resist the change, and what special tactics need to be put into place to mitigate the resistance?
Masculinity/Femininity In masculine cultures, tough, decisive, and aggressive leaders are normal. They are more concerned with the economic performance of the organization and are more likely to use coercion and manipulation to deal with resistance to change. Employees are more likely to show anger and resist the change if they feel it is being imposed upon them. In feminine cultures, leaders are more likely to be supportive and to be facilitators for effective cooperation. They tend to be more concerned with getting the best out of people to achieve a common goal, and they switch from the decision-maker role to the mediator role as required. Employees are less likely to strongly resist change because they feel supported. When resistance does occur,	Based on your observations and experience with the organization, ask yourself: • What motivates employees? Is being the best more important to them, or is liking what they do more important? How can this be leveraged to manage resistance to change? • What form is resistance likely to take (i.e., apathetic, passive, active, aggressive)? • What approach are managers most likely to use when employees resist change?

(continues on next page)

Table 8.1 Influence of Culture on Resistance to Change (cont.)

Cultural Dimension	Things to Consider
managers are more likely to use negotiation, participation, and inclusion to deal with it.	
Long-Term/Short-Term Orientation Employees in long-term orientation cultures are more likely to encourage flexibility of the change plan. Adjusting previous decisions to new circumstances is not a problem and is unlikely to result in resistance or stronger resistance to the change. Employees in short-term–orientation cultures are more attached to procedures and rules. As a result, they are likely to resist changes to previous decisions, and changes could result in stronger resistance to or rejection of the change itself.	Based on your observations and experience with the organization, ask yourself: • How comfortable are employees with flexibility in the change management plans? Is resistance more likely to increase if previous decisions need to be revisited due to new circumstances? • What type of change are employees most likely to resist—incremental or transformational? • Is more resistance anticipated for changes that take longer to deliver? If so, what actions can be taken to mitigate the resistance?
Indulgence/Restraint Indulgent cultures place more importance on freedom of speech and personal control. As a result, people in indulgent cultures are more likely to feel optimistic about change, but they will not hesitate to voice their objections to the change if they perceive it will result in the loss of something valuable to them. In restrained cultures, there is a tendency toward a fatalistic attitude, a greater sense of helplessness about personal destiny. As a result, people in restrained cultures may be more apathetic to the change. This dimension is also likely to have an impact on generational differences. The impact of technology on younger generations suggests that they may be less resistant to technology changes than are older generations, but more research is still needed in this area.	Based on your observations and experience with the organization, ask yourself: • Do employees generally have an optimistic or a pessimistic outlook on life? How might this influence their reaction to change? • Are there certain types of change that people are less likely or more likely to resist? • Are there large generational differences in the employee population? If so, do you anticipate resistance based on generational variables as well as cultural variables? What tactics can you use to deal with this?

Table 8.2 Comparison of Finland and China Based on Hofstede's Cultural Dimensions: Implications for Resistance Management

Finland	China
• Low power distance • Individualistic • Uncertainty avoiding • Feminine • Short-term oriented (normative) • Indulgent	• High power distance • Collectivistic • Uncertainty accepting • Masculine • Long-term oriented (pragmatic) • Restrained
Finnish Profile—Perceptions and Reactions to Change	**Chinese Profile—Perceptions and Reactions to Change**
Finnish companies tend to have more egalitarian (bottom-up and inclusive) structures. Employees are more likely to expect to be involved in decision making and are more willing to change decisions. As a result, professional disagreements are a more common source of resistance to a change. When resistance does occur, managers are more likely to rely on negotiation, participation, and inclusion to deal with it. The Finnish management style tends to be more consultative and inclusive, and communication between managers and employees is taken for granted. This means communication about the change needs to stand out from day-to-day communication, so more time may need to be spent on "selling" the change. Finns rely on good collaborative planning and hard work. In the planning process, Finns tend to give a lot of attention to worst-case scenarios, but there is actually little fatalism in Finnish business.	Chinese companies tend to have strict hierarchical (top-down) structures. Managers make decisions, and employees are expected to implement those decisions unquestioningly. As a result, management is less likely to have to invest a lot of time in "selling" the change, and employees are less likely to strongly resist or reject the change, unless it violates the benefits of groups, group members, or group leaders. Groups can be a source of resistance. Because there is a socially accepted significant power differential, Chinese managers are more likely to use explicit or implicit coercion or manipulation to overcome resistance. Chinese typically have a fatalistic attitude to situations, and this extends to the business environment. Alone, they cannot influence events much, and they prefer to act in harmony with them. This may influence their reactions to change.

Key Points

- One of the most common reasons people resist organizational change, regardless of culture, is because they believe the change will lead to the loss of something valuable to them.
- The reaction to change is influenced by many variables, including national culture, the organizational environment, and the individual's personality.
- While there are predictable, universal sources of resistance, our perception of and reaction to change can be influenced by our culture.
- Resistance to change can be found in all organizations in all countries, but change is more likely to be strongly resisted in countries that are high in power distance, collectivistic, and high in uncertainty avoidance.
- Analyzing the cultural context can help us make sense of employees' reactions to change initiatives, and get a better understanding of the intricate relation between distinct factors of resistance to change.
- Resistance management strategies require the skillful application of a number of strategies, often in different combinations, and choosing the right strategies is context-dependent.

Want to Know More?

There are numerous books on resistance to change, but very few that specifically address the importance and impact of national culture on reactions to change. *Why Culture Matters—An Empirical Study of Working Germans and Mexicans: The Relationship Between National Culture, Resistance to Change and Communication* by Sonja Schultz (2009) draws attention to the impact national culture has universally, and demonstrates how resistance sometimes arises from cultural insensitivity when implementing change.

Although the study focuses on German and Mexican cross-cultural interactions, it raises awareness of how cultural backgrounds influence reactions to change and the need for individual approaches for solving differences arising from national culture.

Chapter 9

Culture's Influences Vary by Context

The international manager reconciles cultural dilemmas.[*]

— Fons Trompenaars

Observing how a person behaves in one situation is not necessarily indicative of how they will behave in another. Culturally intelligent people are able to compensate for their cultural conditioning when they find themselves operating in another culture by adopting the behaviors they begin to see as appropriate to that culture.

9.1 Managing in Two Cultures

A few years ago, *Strategic Change* published an interesting study on the roles of Chinese managers involved in organizational change in foreign-owned enterprises (FOEs) in China (Kong and Gao, 2009). The Chinese managers who took part in the study were drawn from the top elite in terms of academic qualifications, managerial know-how, progressive thinking, and rank. Despite differences in backgrounds and gender, it was common among the Chinese managers

[*] https://quotefancy.com/quote/1758303/Fons-Trompenaars-The-international-manager-reconciles-cultural-dilemmas

in the study that they began working in foreign companies right after graduation from university, and none had experience working in Chinese state-owned enterprises. Due to their exposure to cross-cultural working environments and international business circles, these managers were under greater pressure and were faced with significantly more challenges than their peer Chinese managers who did not work for FOEs.

For example, they spent a lot of time going back and forth between Chinese staff and foreign managers to promote a harmonious environment in the company. The Chinese managers regarded the complexity of cross-cultural management as an opportunity, because foreign managers often felt overwhelmed by the various peculiarities of Chinese culture.

The study identified three roles the managers played in organizational change: change agent, change follower, and anti-change agent.

- **Change agent.** Chinese managers in this study played two important change agency roles: cross-cultural management and conflict resolution—helping Chinese staff get used to Western management styles and helping foreign managers learn to build trust with and manage Chinese workers. With regard to conflict resolution, the Chinese managers used flexible strategies, which reflected their capability in dealing with different cultures. The research also found that Chinese managers were more likely to play a change agent role when they kept a close relationship with the foreign managers.
- **Change follower.** Sometimes even very high-level Chinese managers hesitated to get involved because they could not cope with the complications of change-related initiatives. As a result, they became change followers rather than change agents. Also, for some of the Chinese managers, being a change follower was simply a safer choice.
- **Anti-change agent.** Some of the Chinese managers showed negative emotions toward change in some situations, especially when they perceived that the changes were not in the common interest of a group (in-group) or negatively impacted things such as their position or compensation. According to the researchers, Chinese managers often link their career development with organizational change and try to benefit from the change in order to gain more power and status. (Only two Chinese managers in the study assumed a position of anti-change agent, because such a role requires more power and a higher position.)

The study further identified six factors that influenced which of the three roles participants were likely to assume based on certain contextual factors (see Table 9.1).

Table 9.1 Factors Influencing the Roles of Chinese Managers in Organizational Change in FOEs

Factor	Findings
Type of foreign-owned enterprise (FOE): Joint venture or wholly foreign owned	• Complexity and flexibility of Chinese managers' roles varied according to the type of company in which they worked. • Chinese managers in wholly foreign-owned enterprises played more complex and flexible roles than those in Chinese/foreign equity joint ventures.
Nationality of FOE: American, German, Japanese, Korean	• Nationality of the foreign investors determined the general organizational culture in FOEs. • Chinese managers working in American companies experienced the greatest freedom to express their ideas and get involved in organizational changes and, as a result, exhibited the most optimistic reactions to change. • Chinese managers working in German companies exhibited the most negative reactions to change. • Situations in the Japanese and Korean companies fell somewhere in between that of the American and German companies. • Overall, the greater sense of achievement the Chinese managers felt, the more positive attitude they had to the organization change they led.
Degree to which Chinese managers are socialized by the company	• Chinese managers who were not socialized very much within the FOE were more likely to be change followers rather than change agents, because they did not have the understanding of the firm and trust by their foreign bosses to influence change.
Degree of factionalism impact within the company (degree of difference between in-group and out-group sentiment)	• Chinese managers were likely to see themselves as similar to the other Chinese managers and different from foreign managers • Factionalism still existed among Chinese people themselves, and the level of factionalism directly influenced the power of Chinese managers individually or as a group by increasing or reducing the importance of the Chinese managers in organizational change. • The higher the level of factionalism, the less important a Chinese manager was in organizational changes. Accordingly, if Chinese managers wanted to increase their importance in organizational changes, there was no choice but to reduce factionalism.

(continues on next page)

Table 9.1 Factors Influencing the Roles of Chinese Managers in Organizational Change in FOEs (cont.)

Factor	Findings
	• Although these managers could be trained, many more solutions lay in how to build an organization in which working relationships, not interpersonal relationships, dominated. • In wholly foreign-owned companies, the situation was more complex, because added to the individual differences in age, gender, and class, there were few shared goals and common backgrounds among the Chinese managers.
Interpersonal relationship between Chinese and foreign managers	• The closer the relationship between Chinese managers and their foreign bosses, the more active roles the Chinese managers played as change agents or good change followers, because they were given trust and motivation to conduct the change. • Those who had less of a personal relationship or no personal relationship with the foreign managers treated organizational change as a threat.
Gender	• Female Chinese managers generally showed different attitudes and behaviors to organizational change, as compared with their male counterparts. They cared more about job stability, were more easily satisfied, and encouraged participation and information sharing during changes. • Male Chinese managers were more likely to use a direct command-and-control leadership style. (It cannot be generalized from this study that female Chinese managers were more likely to be change followers rather than change agents or anti-change agents.)

Findings in the research suggest that the Chinese managers were reluctant to take risks to force through their individual decisions. On the one hand, they believed that traditional Chinese leadership values, such as working for the greater benefit of employees, still resulted in effective leadership and job commitment among Chinese staff; thus, they showed such leadership style in their interaction with Chinese staff. On the other hand, they tried to adjust themselves to outwardly reflect a style that was closer to that of their foreign managers, because they perceived that it was necessary for them to do so in order to survive and get promoted. And they were the first group of people to effectively communicate foreign management culture and norms to other Chinese staff, all the while bringing special Chinese traditions to the attention of foreign managers. They tried to combine the best of Chinese and foreign cultures and apply them with flexibility to bring about organizational change.

However, the factionalism that arose from their power struggles prevented them from dealing effectively with various conflicts, because these situations demanded that they draw a clear line between in-groups. And although they had outstanding learning ability during the implementation of changes, they often adopted what they learned superficially and unconditionally in order to be socialized by the organization as soon as possible. But when the next transformational change came, they often got into trouble, because they found that they themselves were resistant to this change.

Research into China's new generation of managers—those having grown up during the era of social reform starting in 1977—suggests that Chinese managers tend to maintain a relatively high level of traditional Confucian values as well as collectivistic tendencies (Ralston et al., 1995, 1999). And the fact that they greatly value interpersonal relationships conforms with the cultural research findings that the Chinese rank the ability to maintain harmonious relationships in the workplace as the primary reason for their success. But when compared to the previous generation of Chinese managers, the new generation of Chinese managers tends to have a higher individualistic tendency and to act more independently (Ralston et al., 1999).

This study focused on the behavior and thinking of a group of Chinese managers who have spent most or all of the career in FOEs and who play a unique role in striking a balance between China's culture and value systems and those of the West. Due to the cultural distance between China and the West, the challenges that these managers faced may be greater than those faced by someone managing change in an environment in which the cultures are not as distant. Nevertheless, it is important to recognize national culture as an important contextual variable, even when the cultures are more similar than dissimilar.

In global change initiatives, you are likely to simultaneously encounter multiple cultures, both close and familiar as well as distant and exotic. Consider how you would approach each of the following situations. Thinking back to the cultural dimensions we have reviewed, what things would you want to understand and take into consideration when planning your change-management interventions in each situation?

- You are a Dane managing an organizational change for a Danish company in Denmark. Your cultural values are likely to be the same as those embraced by both the company and the country.

- You are a Canadian managing an organizational change for a British company in Saudi Arabia. Your cultural values are probably similar to those embraced by the company but very different from the national culture in which you are working.
- You are an American managing an organizational change for a Korean-owned company in the United States. Your cultural values are the same as the national cultural but are likely to be dissimilar to the values embraced by the company.
- You are a Pole managing an organizational change for an American company in Brazil. Your cultural values are likely to be dissimilar to both the national cultural values and the values embraced by the company.

These scenarios highlight just a few of the endless variations of cultural contexts that confront us when working globally, and this is why national culture has such far-reaching implications for change management. Group orientation in collectivistic countries, for example, means that information is likely to spread quickly within the group, it is easier to achieve consistent behavior among group members, and cooperation between group members tends to be more effective. In individualistic countries, on the other hand, resistance to change is more subject to individual response than to group reaction. What are the implications for WIIFM ("What's in it for me")? Traditional change management theories place a lot of emphasis on communicating facts, such as the need for change, the benefits of the change, and the risks of not changing. This may have greater appeal in individualistic countries, but in more collectivistic countries change leaders and managers may need to pay more attention to employees' emotions and feelings. Resistance can often be reduced if leaders communicate messages in a way that evokes employees' emotional response.

Many researchers (Adler, 1997; Hofstede, 1980, 1993; Schneider and Barsoux, 2003) have questioned the cross-cultural transferability and applicability of management practice, including change management interventions. Most change management models have been developed in highly individualistic and low power distance cultures based on cultural assumptions that may not hold in every cultural context. If you are a change agent operating across national boundaries, you need reliable information about national culture that can give you insights that will help you develop more culturally sensitive change management strategies. There is a wealth of research available to you—although sometimes overlapping and contradictory—on cross-cultural difference and the impact of cultural differences on organizational behavior. The best known is Hofstede's (1980, 2000) model of cultural dimensions, but there are many others.

Again, cultural dimensions are only one way of understanding a very complex world. They can help us understand that what happens in one culture does not necessarily happen in another. And it is important to remember that the cultural-dimensions theory is about cultural groups and not individuals, who can vary widely despite sharing a culture.

Think back to the nested nature of culture (refer to Figure 2.1, page 13). While the individual reflects the values of the various layers of culture surrounding them, they have their own unique personality. And people are able to compensate for their cultural conditioning when they find themselves operating in another culture by adopting the behaviors they begin to see as appropriate to that culture, as we saw with the Chinese managers working in foreign-owned enterprises. Cultural differences, along with all of the other contextual variables (age, gender, profession, etc.) should influence the appropriateness of any change management strategy. Context matters.

Do you use the same change management interventions regardless of the cultural setting? It is tempting to assume that if something worked in one place, it will in another. But as the practice of change management grows, it is increasingly important for us to refine it to take national culture into account so that our change management interventions will be culturally mindful and appropriate for the cultural context. Cultural sensitivity is one of the key traits of successful international managers. And when it comes to managing change across cultures, your success lies in your ability to understand and adapt to different cultural contexts while still preserving your core cultural and individual values.

- What are your own managerial values? Have they been changed by your exposure to cross-cultural environments? If so, to what extent?
- How do you sell your vision or ideas to foreign subordinates and managers? How do you secure their trust, cooperation, commitment, and participation? How do you reconcile the cultural differences?
- What are some of the unique challenges you face when managing change in a global environment as compared to when you are managing change in your home culture?
- Which cultural dimensions or specific values stand out for you as providing insight or practical understanding of your own culture or those you come in contact with in your change management work?

Key Points

- In global change initiatives, you are likely to encounter simultaneously multiple cultures, both close and distant.
- Culturally intelligent people are able to compensate for their cultural conditioning when they find themselves operating in another culture by adopting the behaviors they begin to see as appropriate to that culture.
- There are endless variations of cultural contexts when working globally, and this is why national culture has such far-reaching implications for change management.
- Cultural differences, along with all of the other other contextual variables that are unique to the situation, should influence the appropriateness of any change management strategy. Context matters.

Want to Know More?

If you want to learn more about contextual intelligence, *Contextual Intelligence: How Thinking in 3D Can Help Resolve Complexity, Uncertainty and Ambiguity* by Matthew Kutz offers a structured framework for critical thinking and decision making that shows how to use hindsight, insight, and foresight to navigate through complexity (Kutz, 2017).

Chapter 10

Key Competencies for Leading and Managing Change Across Cultures

*In addition to extraordinary business leadership skills, a leader now needs cultural intelligence. Cultural intelligence requires transcending one's own cultural background to interact with diverse and unknown intelligences.**

— E. S. Wibbeke

Change management is a context-sensitive exercise that needs to be culturally adapted to its environment. Leading and managing change across cultures requires an investigation of not only the context of the situation, but also the cultural dynamics that are at play. It requires cognitive ability (intellectual clarity), emotional sensitivity, and an ability to adapt different leadership styles in order to effectively influence people from other cultural backgrounds. Cultural intelligence (CQ) (Ang et al., 2007) and the global mindset (GM) (Javidan and Teagarden, 2011) are two streams of research that attempt to find a personality type or mindset that might be correlated with superior performance among successful domestic and international managers.

* Wibbeke, E. S., and McArthur, S. (2013)

10.1 Cultural Intelligence

Cross-cultural interactions require people to understand and adapt their behaviors to the values, beliefs, and customs of other societies in order to promote more effective interactions and relationships (Ng, Tan, and Ang, 2011). The concept of CQ has emerged as both an important theoretical framework for identifying intercultural competence and a practical competency for anyone working in a global environment (Ang, Van Dyne, and Rockstuhl, 2015). CQ focuses on our ability to successfully adapt to new cultural contexts and function effectively within a cross-cultural environment.

Being culturally intelligent isn't just about having good general social skills. CQ is a specific form of intelligence focused on our capabilities to grasp, reason, and behave appropriately and effectively in situations characterized by cultural diversity. It enables us to appreciate the diversity of experiences and to formulate rapid, accurate, and contextually sensitive responses to emerging issues. In some ways, CQ is similar to emotional intelligence or emotional quotient (EQ), which is another important competency for change leaders. But emotional intelligence itself cannot be meaningfully understood outside of its cultural context, and it has different effects on management outcomes in different cultural contexts. How does national culture influence the EQ of individuals?

Culture determines the values and norms of individuals. What is considered important in a society is determined to a great degree by culture. Why we feel an emotion may be different than why someone from another culture feels an emotion. Societal norms determine the meaning of emotions and the controlling of them (Eid and Diener, 2001). The kinds of emotions that people openly show and how those emotions are communicated differ widely across cultures (Matsumoto, 1989). Emotionally intelligent individuals are able to code and decode their own and others' emotions as they are displayed in their own culture, but this becomes a much more difficult task in different cultural settings, especially when the cultures are dissimilar.

Cultural intelligence predicts a variety of important outcomes in the workplace, such as cross-cultural judgment and decision making, cultural adjustment, idea sharing, and job performance (Ramsey and Lorenz, 2016). Employees who possess a high level of cultural intelligence can play an important role in bridging divides and knowledge gaps in an organization because of their ability to help build interpersonal connections, educate peers about different cultures, integrate diverse resources, and make the best use of multicultural perspectives and approaches.

Researchers (Earley and Ang, 2003) found that our capacity to adapt to unfamiliar cultural environments is based on four dominant aspects: motivational, cognitive, metacognitive, and behavioral (Figure 10.1).

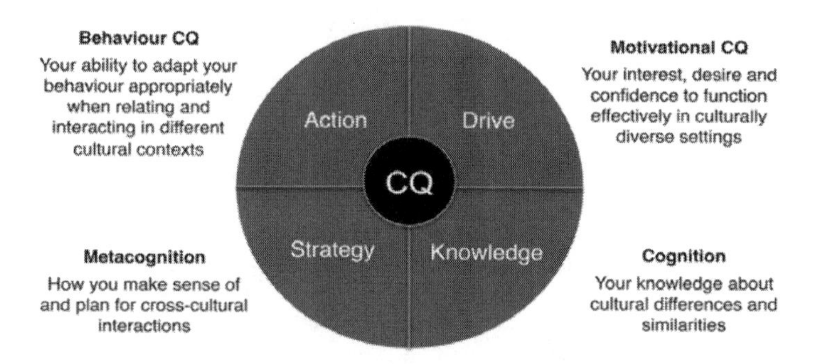

Figure 10.1 Four-Factor Model of Cultural Intelligence (*Source:* Based on data drawn from Earley, P., and Ang, S. 2003. *Cultural Intelligence: Individual Interactions Across Cultures,* Vo1. 1. Stanford, CA: Stanford Business Books)

10.1.1 Motivational CQ (Drive)

Motivational CQ reflects our openness and willingness to engage in cultural interactions and our perseverance and tenacity to continue engaging in them even when we experience failures and setbacks. This is a key component in activating the cognitive aspect of CQ (Ang, Van Dyne, and Tan, 2011), and there is also correlation to cultural psychological capital, which in turn relates to increased levels of metacognitive awareness (Figure 10.2).

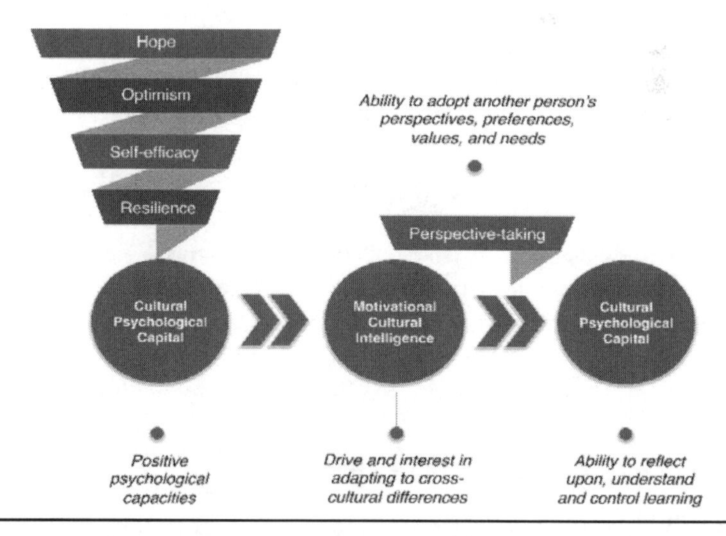

Figure 10.2 Linkage Between Psychological Capital, Motivational CQ, and Metacognitive CQ (*Source:* Adapted with permission from Yunlu, D. G., and Clapp-Smith, R. 2014. Metacognition, Cultural Psychological Capital and Motivational Cultural Intelligence. *Cross Cultural Management,* 21(4): 386–399)

Cultural psychological capital is composed of four sub-dimensions: *hope, optimism, self-efficacy,* and *resilience* (Luthans et al., 2007; Yunlu and Clapp-Smith, 2014) (see Table 10.1).

Table 10.1 Sub-Dimensions of Cultural Psychological Capital

Hope	Persevering toward goals and redirecting paths to goals when necessary in order to succeed in a cross-cultural environment
Optimism	Making a positive presupposition about succeeding now and in the future in an international context
Self-Efficacy	Having confidence to take on and make the required effort to succeed at challenging tasks in cross-cultural settings
Resilience	Attaining success by sustaining and bouncing back when faced by problems and roadblocks in an international setting (Luthans et al, 2007, p.3)

Individuals with high motivational CQ have an intrinsic interest in learning in cross-cultural situations. Hope, optimism, self-efficacy, and resilience are considered to be intrinsic motivational propensities (Luthans et al., 2007).

A recent study (Yunlu and Clapp-Smith, 2014) provides an interesting glimpse into some of the cognitive processes that are relevant to learning in cross-cultural contexts. It highlights the relevance and importance of psychological resources, not just for multinational corporations but also for the broader population of managers, particularly as the economy becomes more globalized and diverse populations become more mobile. Research has also shown that positive psychological capital can be developed through learning interventions (Luthans et al., 2006).

10.1.2 Cognitive CQ (Knowledge)

The cognitive aspect of CQ is based on our self-awareness—awareness of our own culture, personality, leadership style, etc. But a high level of self-awareness alone does not guarantee that we will be more effective in cross-cultural interactions. We must have a desire to learn, to be curious, and to engage and re-engage despite uncertainty, setbacks, or failure in our cross-cultural encounters.

10.1.3 Metacognitive CQ (Strategy)

Metacognitive awareness deals with the knowledge of cognitions (thoughts and perceptions) as well as the regulation of cognitions. In other words, it is our ability to acquire and understand cultural knowledge, which is associated with our ability to re-examine our cultural assumptions and adjust our mental model accordingly—that is, to take unfamiliar patterns of behavior and inductively create a proper mapping of social situations in order to function more effectively in unfamiliar situations. When we have high metacognitive cultural intelligence, we are able to more readily reflect upon interactions and better adjust our cultural knowledge in multicultural situations (Ang and Inkpen, 2008; Earley and Ang, 2003).

There is a correlation between metacognition, motivational cultural intelligence, and learning. Having an ability to understand and regulate how we learn provides greater opportunities for us to prepare ourselves for challenging work contexts, including the complexity of cross-cultural work environments. And by understanding the role of motivational cultural intelligence in the recognition and regulation of thoughts and perceptions, change managers are better able to make a case for and encourage organizations to focus greater formal or informal resources on encouraging learning from different cultural contexts. And because motivational cultural intelligence is an intrinsic motivation in directing energy toward adapting to new environments, organizations may find avenues to reinforce and promote ability by creating conditions to support motivational cultural intelligence.

10.1.4 Behavioral CQ (Action)

The behavioral aspect of cultural intelligence reflects our ability to adapt our actions and to act appropriately based on the cultural context of the situation—that is, our ability to exhibit appropriate verbal and nonverbal actions when we are interacting with people from different cultures. This involves having a wide and flexible repertoire of behaviors.

People with high behavioral cultural intelligence are better able to exhibit contextually appropriate behaviors based on a broad range of verbal and nonverbal capabilities, such as exhibiting culturally appropriate words, tone, gestures, and facial expressions (Hall, 1959).

Cultural intelligence really is a skill that we can't do without when leading change in today's global economy. The good news is that anyone who is motivated enough can cultivate cultural intelligence—but it takes practice. As the old Finnish expression reminds us, "No one is a blacksmith when they are born" ("*Ei kukaan ole seppä syntyessään*"). That is, no one can be expected to be an expert at something before they have had the chance to practice it.

10.2 Cultural Intelligence

The process of learning to recognize and understand cultural differences often starts by our unintentionally not paying any attention to them at all. By default, we interpret different situations from the perspective of our own cultural framework, and we unconsciously assume that other people are interpreting those situations in a similar way. When this happens, cross-cultural clashes and conflicts can begin to emerge. Cultural intelligence encompasses a set of knowledge, skills, and abilities that help us to better understand how individuals think and behave in different global settings. So, if CQ is important, how can it be developed?

Research has shown that CQ can be trained and developed through exposure to the inherent differences arising from cross-cultural interactions (Ng et al., 2011). With each new interaction, our mental schemas are being refined and broadened to accept additional inputs that can, in turn, act as prompts when we face new uncertain and ambiguous encounters. CQ and cross-cultural interactions—whether in our own country or abroad—are mutually self-reinforcing and interdependent. In this way, our interactions with people from different cultures become more than the sum of their parts.

Here are six powerful things you can do to become more culturally intelligent:

- **Become self-aware.** Before you can begin to understand the cultures of other people, you need to understand your own. Recognize your own cultural values, beliefs, assumptions, and personal biases. By doing this, you'll be able to think about how these traits might impact your approach to and understanding of differences.
- **Be curious.** Ask questions. Learn about the history, traditions, and values of different cultures. Although you can't learn everything about every culture, you can certainly gain useful insights into the workplace and business expectations in other countries and markets.
- **Be prepared.** Get to know the differences and similarities between your culture and the cultures of others. Learn how business is conducted in relevant countries and markets. Find out what factors may influence how people in different cultures perceive and react to change.
- **Build strong intercultural relationships.** Forge relationships with people from other cultures. Developing relationships with people from other cultures facilitates valuable learning. The more you interact with people from different cultures, the more confidence you'll have in different cultural contexts, and the less you'll resort to stereotyping.
- **Develop strategies to adjust your style**. Learn to appreciate diversity and don't work against it. Be flexible—learn to behave in ways that are appropriate to the cultural context, even if doing so tests your own abilities and pushes you outside of your personal comfort zone. Find a mentor

who can help you appropriately adjust and adapt your style to the cultural context.

- **Be more accepting.** Develop an open mind. People from other cultures may have different views, habits, and interests, and being open to these differences can make you more approachable.

Much cross-cultural interaction begins as a trial-and-error process. Some of your efforts will be more successful than others, but every cross-cultural interaction is an opportunity for you to learn and to continue to build your repertoire. Placing knowledge into practice is an essential way to develop cultural intelligence.

10.3 Global Mindset

How do you feel about people, places, and things that are foreign to you? Many managers from developed (WEIRD) countries approach foreign markets with a sense of arrogance or superiority—consciously or unconsciously. A mindset is the mental attitude that determines how we perceive and respond to situations. Having a global mindset is about being comfortable with being uncomfortable in unfamiliar environments, stepping outside of our comfort zone and showing our curiosity about new things, new ideas, and new concepts. Global mindset is a combination of awareness and openness to the diversity of cultures and markets combined with a desire and capability to integrate across the diversity. It is an ever-developing process built upon cognitive feedback mechanisms that encourage the search for life experiences that expand and refine an individual's mental schemas (Gupta and Govindarjan, 2002).

Developing a global mindset requires more than just studying international business and travelling. Even a well-travelled person can still be blissfully ignorant of cultural nuances and hopelessly narrow minded. Global mindset requires keen observation of how foreigners interact with each other, recognizing how their culture, politics, economy, religion, etc. influence their behavior and how they contrast with our own, and then leveraging and using that awareness when conducting business with people from other cultures. Really, having a global mindset is beneficial for any employee in any company (Gupta and Govindarjan, 2002).

Global mindset comprises three components, or "capitals"—*intellectual capital, psychological capital,* and *social capital*—and each capital is underpinned by three corresponding building blocks or attributes (Javidan and Teagarden, 2011) (Figure 10.3).

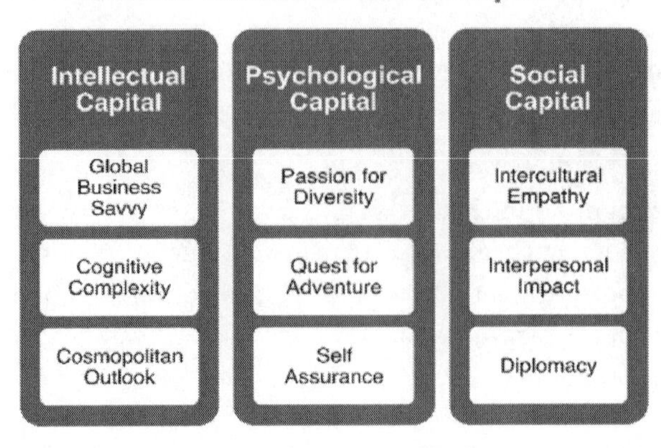

Figure 10.3 Global Mindset Core Capitals and Underpinning Building Blocks (Data drawn from Javidan, M., and Teagarden, M. B. 2011. Conceptualizing and Measuring Global Mindset. *Advances in Global Leadership*, 6)

10.3.1 Intellectual Capital

Intellectual capital encompasses an individual's knowledge and cognitive capabilities regarding different cultural contexts. It involves having a good grasp of global issues, being cognitive of interdependencies in the global landscape, understanding cultural differences, and being able to manage complex cultural issues. And it also involves recognizing that sometimes the business approaches we use successfully in our own culture (change management approaches, for example) may not work wholesale in other cultural contexts.

Intellectual capital is underpinned by three key intellectual attributes:

- **Global business savvy.** Knowledge of the way business is conducted in different parts of the world including business strategies, risk management, supplier options, etc.
- **Cosmopolitan outlook**. Understanding that things can be done differently in different parts of the world; knowledge of geographies, histories, economic and political issues, important world events, etc.
- **Cognitive complexity**. Ability to grasp, digest, interpret, and leverage large amounts of information, including complex concepts and abstract ideas.

Intellectual capital is by far the easiest to develop. You can develop it by reaching and becoming more aware of cultural differences, cross-cultural issues, and global events; by participating in cultural activities and global professional organizations; and by reading books and local (foreign-language) newspapers. A quick search of the internet will turn results for a plethora of sites where you can find in-depth reports on countries, scholarly research on cultural dimensions, and tools to help you better understand your own culture as well as the culture of others.

10.3.2 Psychological Capital

Psychological capital refers to a positive psychological profile and personality traits. It requires adaptability, self-confidence, resiliency, and optimism. Psychological capital also plays a role in the motivational CQ process, which in turn relates to metacognitive awareness gained from cross-cultural experiences.

Psychological capital is underpinned by three psychological attributes: *passion for diversity, quest for adventure,* and *self-assuredness.*

- **Passion for diversity.** Curiosity about and joy in dealing with people from different cultures; interest in the customs and perspectives of people from different parts of the world
- **Quest for adventure.** Willingness to engage in cross-cultural interactions, to deal with unfamiliar and challenging situations, to take risks, and test your own abilities
- **Self-assuredness.** Having the energy and confidence to take on global assignments and to be comfortable with unfamiliar and uncomfortable situations

Psychological capital is by far the most difficult to develop, because there are limits to how much you can or should try to change your own personality. But you can make improvements in this area by deeply reflecting on your own culture, values, beliefs, assumptions, and biases, and then taking stock of whether or not, and why or why not, you need to make any changes. And you can seek out and expose yourself to new cultural experiences and ideas.

10.3.3 Social Capital

Social capital focuses on internal and external relationships, including an individual's intercultural empathy, interpersonal impact, and diplomatic skills and ability. It is underpinned by three key social attributes:

- **Intercultural empathy.** Ability to connect, communicate, and collaborate with people from other cultures; ability to understand the nonverbal expressions of people from other cultures
- **Interpersonal impact.** Ability to negotiate and build credibility with people from other cultures; ability to build personal and professional networks of influence across borders

- **Diplomacy.** Ability to make a positive impression on people from other cultures; openness to what people from other cultures have to say; willingness to collaborate

Social capital is largely relationship based, so you can develop it by increasing your cross-cultural interactions. Widen your circle of interactions to include more people whose culture and ideas are different from your own. Greater cultural exposure has been linked to greater cultural intelligence (Crowne, 2013) and a better ability to cognitively manage the demands of multiple cultures (Dragoni and McAlpine, 2012). And exposure to more culturally distant cultures has also been linked to stronger strategic thinking competency (Dragoni et al., 2014).

A global mindset is about thinking globally, which is critical to operating in a global environment and even in a domestic environment. Professionals with a global mindset leverage all that they know about their culture and the cultures of other people to react to situations in the most productive ways, all without losing sight of who they are. When you think globally, you are able to develop better communications, relationships, and understanding among colleagues, customers, and global partners.

Globalization has made the business environment more complex, dynamic, and competitive. To understand how people perceive and respond change, you need to be able to look beyond your own cultural boundaries and preconceptions, and this requires a good dose of cultural intelligence and a global mindset. But a global mindset is a behavioral transformation that doesn't happen overnight. Considerable practice is involved in developing these skills, which enable you to become a conduit between cultures. Placing knowledge into practice is the essential way to develop these skills.

10.4 CQ and Global Mindset in Practice

In the November 2016 issue of the *Harvard Business Review,* SAP's CEO, Bill McDermott, recounted his journey to becoming the first American head of a German multinational, and the unique challenges of leading a company that is rooted in a foreign culture. McDermott was raised in a working-class family in the cultural melting pot of Long Island, New York. Although he wasn't well travelled in his early life, he worked in a variety of jobs in which he had to deal with all kinds of people, and this experience taught him a lot about diversity and empathy for other people.

At the age of 29, while working as a sales manager for Xerox, McDermott received his first overseas assignment when he was charged with turning around the company's failing business in Puerto Rico. He wasn't familiar with either the market or the culture, so, rather than arriving with a preset agenda, he spent the first few weeks just meeting with and listening to people in order to understand why some of them were performing poorly. And because he didn't speak Spanish, he asked his assistant to give him phonetic spellings of important Spanish phrases so he could better relate to his new team. His efforts were appreciated, and together he and his team managed to turn the performance of the operation around.

Coming up through the sales function, McDermott learned to adapt his business strategies to different markets and to be sensitive to the nuances of the unique cultural dynamics from country to country. In New York, he learned to be concise and to the point and to focus on the sell. In Asia and other parts of the world, he found that he needed to work more slowly, focusing more on the relationship than on the transaction.

In 2002, SAP asked McDermott to become chief executive of its struggling North American business. In McDermott's view, part of the reason SAP was struggling in North American was down to its leaders' mistakenly assuming they could simply transfer the strategies that worked well in Germany (and other established markets) to the US market. And as he began travelling to Germany to meet with the company's executive committee, McDermott became more acutely aware of the differences between Americans and Germans. He found that German sales and management styles did not fit with American styles and vice versa, and he learned that he had to adapt his own style to suit the cultural context. As he puts it, "Leading in any country is all about reading the room, respecting the culture, and understanding the nuances of how people perceive information. You have to care about what the culture needs instead of just focusing on your agenda and how to get it done" (McDermott, 2016).

In 2014, McDermott became the CEO of SAP. Although he kept his house in Philadelphia, he made the decision to move to Heidelberg, because he felt is was symbolically and strategically important for him to have a residence in Germany. So, what advice would McDermott give to someone who is asked to lead a company based outside of their home culture? Learn to read the room, understand and respect the dynamics of the culture, be empathetic, and give people a compelling vision.

Given the hyper-connectivity of the world today, can you be an effective change leader without developing and exercising a global mindset? How might cultural intelligence help you to be more robust in your change management interventions?

Key Points

- Leading change across cultures requires cognitive ability (intellectual clarity), emotional sensitivity, and an ability to adapt different leadership styles in order to effectively influence people with other cultural backgrounds.
- Cultural intelligence (CQ) and global mindset (GM) are two streams of research concerned with finding a personality type or mindset that correlates with superior performance among domestic and international leaders.
- CQ is an important theoretical framework for identifying intercultural competence and a practical competency focused on capabilities required to grasp, reason, and behave appropriately and effectively in culturally diverse situations.
- CQ comprises four components: *motivational* (drive), *cognitive* (knowledge), *metacognitive* (strategy), and *behavioral* (action).
- GM is characterized as a combination of awareness and openness to the diversity of cultures and markets combined with a desire and capability to integrate across the diversity.
- GM comprises three components or "capitals": *intellectual capital, psychological capital,* and *social capital.*
- Anyone who is motivated enough can develop CQ and GM, but it takes time and practice.

Want to Know More?

You can learn more about cultural intelligence by reading HBR's 10 Must Reads on Managing Across Cultures with a featured article, "Cultural Intelligence" by P. Christopher Earley and Elaine Mosakowski (Earley and Mosakowski, 2004).

To learn more about how to adapt your behavior across cultures while staying authentic and grounded in your own natural style, you might like to read Andy Molinsky's insightful book *Global Dexterity* (Molinsky, 2013).

To learn more about global mindset, read *Developing Your Global Mindset: The Handbook for Successful Global Leaders* by Mansour Javidan and Jennie Walker (Javidan and Walker, 2013).

You can also read Bill McDermott's article "SAP's CEO on Being the American Head of a German Multinational" in the November 2016 issue of the *Harvard Business Review* (McDermott, 2016).

Chapter 11

Conclusion

As the practice of change management grows, it is increasingly important for us to refine its application. Most interventions to organizational change have been developed in highly individualistic and low power distance cultures. Yet, it is often tempting to fall back on the same approach regardless of the cultural setting, assuming that if it worked in one place, it will in another. Some interventions transfer intact, many do not. Don't give in to the temptation. Don't assume that your preferred approach to change management will always be compatible and adequate in cultures that are different from your own. Learn to understand and respect the differences between ways of working in different cultural contexts. The consequences of getting things wrong can be dramatic.

If you are a change agent operating across national boundaries, you need to take national culture into consideration as an important contextual variable when developing your change management strategies. The first step in designing a culturally compatible change effort is to have an understanding of how cultures vary. In this book we've looked at some of the cultural models that are especially useful to consider when designing change management interventions. All of them have important factors that can provide insightful explanations that will contribute to your understanding of national culture as it relates to management practices and managing change in different cultural contexts. Of course, making assumptions about individuals based solely on where they are from is risky and ill advised, but appreciating how culture shapes behavior and perception is essential preparation for working across cultures. The cultural frameworks provide a good starting point for planning your change strategy.

In global change efforts, you are likely to encounter multiple cultures, both close and familiar as well as distant and exotic. Try not to view cross-cultural situations through your own cultural lens. Use evidence from the various cultural models to predict and prevent cultural misunderstandings and to better prepare yourself to cope with the challenges of change agency in different cultural settings. Be observant and recognize differences in ways of working and perceptions and reactions to change in different cultural contexts. Learn to use the information gathered through your research and observations to examine your own culture, behaviors, and ways of working in order to rise above and bridge cultural differences.

Managing change across cultures is about understanding comfort zones and respecting differences. Don't expect business to be the same across all cultures, and don't assume that if something is good for you it is good for your foreign colleague, or that their mind-set is the same as yours. Seek understanding of how and why differences exist rather than simply benchmarking their behaviors and ways of working against your own. Learn to appreciate the subtleties of communication and to accord an appropriate amount of time when important decisions need to be made. Some cultures restrain people from immediate, outright agreement or disagreement on important issues. And don't underestimate the importance of relationship building and trust. In some cultures, people are unlikely to want to deal with you or to be open with you until you have spent some time with them. Ask colleagues to coach you on the nuances and niceties of their culture, and coach them on yours in return.

Cultural sensitivity is one of the key traits of successful international managers, so learn to approach your change efforts in a culturally mindful way. Fine-tuning your cultural intelligence is a never-ending journey that requires a deep-seated curiosity about countries, cultures, and traditions that are different from your own. Curiosity does not come naturally to everyone. It requires effort. Maybe you are never going to learn about every culture in depth, but raising your own awareness of how culture influences behaviors in the workplace and perceptions of and reactions to change can pay big dividends in your cross-cultural change efforts. Even something as small as being able to introduce yourself and exchange pleasantries in the local language can buy credibility with colleagues and partners.

When it comes to managing change, national culture matters. But it would be a mistake to treat all of your foreign colleagues and partners as part of a homogenous group or to expect everyone in a given culture to behave in exactly the same ways. Recognize that individuals are not simply products of their culture. They are a mix of many variables—culture, ethnicity, environment, gender, age, personality, profession, etc. When managing change, all of the variables of the situation need to be taken into account. Context matters.

References

Adler, E. 1997. Seizing the Middle Ground: Constructivism in World Politics. *European Journal of International Relations,* 3(3): 319–363.

Allred, L. J., Chia, R. C., Wuensh, K. L., Ren, J. J., and Miao, D. M. 2007. *In-Groups, Out-Groups and Middle-Groups in China and the United States.* La Mesa, CA: National Social Science Association. Retrieved February 28, 2018, from http://www.nssa.us/journals/2007-29-1/2007-29-1-02.htm

Alvesson, M. 1993. Organizations as Rhetoric: Knowledge-Intensive Firms and the Struggle with Ambiguity. *Journal of Management Studies,* 30(6): 997–1015.

ASTD. 2012. *The Global Workplace.* Alexandria, VA: American Society for Training and Development.

Ang, S., and Inkpen, A. 2008. Cultural Intelligence and Offshore Outsourcing Success: A Framework of Firm-Level Intercultural Capability. *Decision Sciences,* 39: 337–358.

Ang, S., Van Dyne, L., and Rockstuhl, T. 2015. Cultural Intelligence: Origins, Conceptualization, Evolution, and Methodological Diversity. *Handbook of Advances in Culture and Psychology,* 45: 141–164.

Ang, S., Van Dyne, L., and Tan, M. 2011. Cultural Intelligence. In R. J. Sternberg and S. B. Kaufman (Eds.), *The Cambridge Handbook of Intelligence,* pp. 582–602. New York, NY: Cambridge University Press.

Ang, S., Van Dyne, L., Koh, C., Ng, K., Templer, K., Tay, C., et al. 2007. Cultural Intelligence: Its Measurement and Effects on Cultural Judgment, Decision Making, Cultural Adaptation and Task Performance. *Management and Organization Review,* 3: 335–371.

Arnett, J. J. 2002. The Psychology of Globalization. *American Psychologist,* 57(10): 774–783.

Aronson, E., Wilson, T. D., and Akert, R. 2010. *Social Psychology* (7th Edition). Upper Saddle River, NJ: Prentice Hall.

Barbaro, M. 2017 (August). *Germany: Wal-Mart Finds That Its Formula Doesn't Work in All Cultures.* Berkeley, CA: CorpWatch. Retrieved March 19, 2018, from http://www.corpwatch.org/article.php?id=13969

Barron, A., and Schneckenberg, D. 2012. A Theoretical Framework for Exploring the Influence of National Culture on Web 2.0 Adoption in Corporate Contexts. *The Electronic Journal of Information Systems Evaluation,* 15(2): 76–186.

Bartunek, J. 1988. The Dynamics of Personal and Organizational Reframing. In R. E. Quinn and K. S. Cameron (Eds.), *Paradox and Transformation: Towards a Theory of Change in Organization and Management.* Cambridge, MA: Ballinger.

Bennis, W. G. 1969. *Organization Development: Its Nature, Origins, and Prospects.* Reading, MA: Addison-Wesley.

Bond, M. H. 1986. Mutual Stereotypes and the Facilitation of Interaction Across Cultural Lines. *International Journal of Intercultural Relations,* 10: 259–276.

Bond, M. H., and Hewstone, M. 1988. Social Identity Theory and the Perception of Intergroup Relations in Hong Kong. *International Journal of Intercultural Relations,* 12: 153–170.

Caldwell, R. 2003. Change Leaders and Change Managers: Different or Complementary? *Leadership and Organization Development Journal,* 24(5): 285–293.

Castro, F. G., Barrera, Jr., M., and Martinez, Jr., C. R. 2004. The Cultural Adaptation of Prevention Interventions: Resolving Tensions Between Fidelity and Fit. *Prevention Science,* 5(1): 41–45.

Chan, A. M., and Rossiter, J. R. 2003. Measurement Issues in Cross Cultural Values Research. *Proceedings of the Australia New Zealand Academy of Marketing Conference,* p. 1586. Adelaide, Australia: University of South Australia.

Coetsee, L. 1999. *From Resistance to Commitment.* Southern Public Administration Education Foundation.

Costigan, R., Insigna, R., Berman, J., Liter, S., Kranas, G., and Kureshov, V. 2007. A Cross-Cultural Study of Supervisory Trust. *International Journal of Manpower,* 27(8): 764–787.

Crawford, L., and Cooke-Davies, T. 2012. *Best Industry Outcomes.* Newtown Square, PA: Project Management Institute, Inc.

Crowne, K. A. 2013. Cultural Exposure, Emotional Intelligence, and Cultural Intelligence: An Exploratory Study. *International Journal of Cross Cultural Management,* 13(1): 5–22.

Dorfman, P., Javidan, M., Hanges, P., Dasmalchian, A., and House, R. 2012. GLOBE: A Twenty Year Journey into the Intriguing World of Culture and Leadership. *Journal of World Business,* 47(4): 504–518.

Dragoni, L., and McAlpine, K. 2012. Leading the Business: The Criticality of Global Leaders' Cognitive Complexity in Setting Strategic Directions. *Industrial and Organizational Psychology,* 5(2): 237–240.

Dragoni, L., Oh, I., Tresluk, P., Moore, O., Van Katwyk, P., and Hazucha, J. 2014. Developing Leaders' Strategic Thinking Through Global Work Experience: Moderating Role of Cultural Distance. *Journal of Applied Psychology,* 99: 867–882.

Du Plessis, Y. 2011. Cultural Intelligence as Managerial Competence. *Alternation,* 18(1): 28–46.

Earley, P. C. 1994. Self or Group? Cultural Effects of Training on Self-Efficacy and Performance. *Administrative Science Quarterly,* 39: 89–117.

Earley, P. C., and Ang, S. 2003. *Cultural Intelligence: Individual Interactions Across Cultures,* Vol. 1. Stanford, CA: Stanford Business Books.

Earley, P. C., and Mosakowski, E. 2004 (October). Cultural Intelligence. *Harvard Business Review.*

Eid, M., and Diener, E. 2001. Norms for Experiencing Emotions in Different Cultures: Inter- and Intranational Differences. *Journal of Personality and Social Psychology,* 81(5): 869–885.

Erez, M., and Gati, E. 2004. A Dynamic, Multi-Level Model of Culture: From the Micro Level of the Individual to the Macro Level of a Global Culture. *Applied Psychology: An International Review,* 53(4): 583–598.

Fiske, J. 2002. *Introduction to Communication Studies.* London, UK: Routledge, Division of Taylor & Francis.

Giebels, E., Oostinga, M. S. D., Taylor, P. J., and Curtis, J. L. 2017. The Cultural Dimension of Uncertainty Avoidance Impacts Police–Civilian Interaction. *Law and Human Behavior,* 41(1): 93–102.

Gupta, A. K., and Govindarajan, V. 2002. Cultivating a Global Mindset. *Academy of Management Executive,* 16(1): 116–126.

Hall, E. T. 1959. *The Silent Language.* Greenwich, CT: Fawcett.

Hall, E. T. 1976. *Beyond Culture.* Oxford, UK: Anchor.

Harzing, A. W., and Hofstede, G. 1996. Planned Change in Organizations: The Influence of National Culture. *Research in the Sociology of Organizations,* 14: 297–340.

Hazel, S. 2016 (February 10). Why Native English Speakers Fail to Be Understood—And Lose Out in Global Business. *The Conversation.* Accessed February 28, 2018, at http://theconversation.com/why-native-english-speakers-fail-to-be-understood-in-english-and-lose-out-in-global-business-54436

Henrich, J., Heine, S. J., and Norenzayan, A. 2010. The Weirdest People in the World? *Behavioral and Brain Sciences,* 33(2–3): 61–83.

Hofstede, G. 1980. *Culture's Consequences: International Differences in Work-Related Values.* Thousand Oaks, CA: Sage Publications.

Hofstede, G. 1993. Cultural Constraints in Management Theories. *Academy of Management Executive,* 7: 81–94.

Hofstede, G. 2001. *Culture's Consequences: Comparing Values, Behaviors, Institutions and Organizations Across Nations,* p. 21. Thousand Oaks, CA: Sage Publications.

Hofstede, G. 2011. Dimensionalizing Cultures: The Hofstede Model in Context. *Online Readings in Psychology and Culture,* 2(8). https://doi.org/10.9707/2307–0919.1014.

Hofstede, G., and Bond, M. 1991. The Confucius Connection: From Cultural Roots to Economic Growth. *Organizational Dynamics,* 16(4): 4–21.

Hofstede, G., and Hofstede, G. J. 2005. *Cultures and Organizations.* New York: McGraw-Hill.

Hofstede, G., Hofstede, G. J., and Minkov, M. 2010. *Cultures and Organizations: Software of the Mind* (3rd Edition). New York: McGraw-Hill Education.

House, R. J., Hanges, P. J., Javidan, M., Dorfman, P. W., and Gupta, V. 2004. *Culture, Leadership, and Organizations: The GLOBE Study of 62 Cultures.* San Francisco, CA: Sage Publications.

Hsu, L. K. 1988. *Americans and Chinese.* Honolulu, HI: University of Hawaii Press.

Hui, C. 1988. Measurement of Individualism–Collectivism. *Journal of Research in Personality,* 22(1): 17–36.

Hui, C. H., and Triandis, H. C. 1986. Individualism and Collectivism: A Study of Cross-Cultural Researchers. *Journal of Cross-Cultural Psychology,* 17: 225–248.

Inglehart, R., and Baker, W. E. 2000. Modernization, Cultural Change, and the Resistance of Traditional Values. *American Sociological Review,* 65(2): 19–51.

Irvine, J. J., and York, D. E. 1995. Learning Styles and Culturally Diverse Students: A Literature Review. In J. A. Banks (Ed.), *Handbook of Research on Multicultural Education,* pp. 484–497. New York: Macmillan.

Javidan, M., and Teagarden, M. B. 2011. Conceptualizing and Measuring Global Mindset. *Advances in Global Leadership,* 6: 13–39.

Javidan, M., and Walker, J. 2013. *Developing Your Global Mindset: The Handbook for Successful Global Leaders.* Edina, MN: Beaver's Pond Press. ISBN-10: 1592989977 | ISBN-13: 978-1592989973.

Kirsch, C., Chelliah, J., and Parry, W. 2011. Drivers of Change: A Contemporary Model. *Journal of Business Strategy,* 32(2): 13–20.

Kluckhohn, F. R., and Strodtbeck, F. L. 1961. *Variations in Value Orientation.* New York: HarperCollins.

Knorr, A., and Arndt, A. 2004 (June 24). Why Did Wal-Mart Fail in Germany? Retrieved February 17, 2017, from http://www.iwim.uni-bremen.de/

Kong, S., and Gao, B. 2009. Chinese Executives in Foreign-Owned Enterprises: Managing in Two Cultures. *Strategic Change: Briefings in Entrepreneurial Finance,* 18(3): 93–109.

Kotter, J. P. 1995 (May–June). Leading Change: Why Transformation Efforts Fail. *Harvard Business Review.*

Kotter, J. P., and Schlesinger, L. A. 1979. Choosing Strategies for Change. *Harvard Business Review,* 57(2): 106–114.

Kouzes, J., and Posner, B. 1993. *Leadership Practices Inventory: A Self-Assessment and Analysis.* Expanded Edition. San Francisco, CA: Jossey-Bass.

Kutz, M. 2017. *Contextual Intelligence: How Thinking in 3D Can Help Resolve Complexity, Uncertainty and Ambiguity.* New York, Shanghai: Palgrave MacMillan. DOI: 10.1007/978-3-319-44998-2. Hardcover ISBN: 978-3-319-44997-5.

Laurent, A. 1986. The Cross-Cultural Puzzle of International Human Resource Management. *Human Resource Management,* 25(1): 91–102.

Lenartowicz, T., and Johnson, J. P. 2007. Staffing Managerial Positions in Emerging Markets: A Cultural Perspective. *International Journal of Emerging Markets,* 2(3): 207–214.

Lewis, R. D. 2006. *When Cultures Collide: Leading Across Cultures.* London, UK: Nicholas Brealey Publishing.

Kong, S., and Gao, B. 2009. Chinese Executives in Foreign-Owned Enterprises: Managing in Two Cultures. *Strategic Change: Briefings in Entrepreneurial Finance,* 18(3): 93–109.

Markus, H. R., and Kitayama, S. 1991. Culture and Self: Implications for Cognition, Emotion and Motivation. *Psychological Review,* 98(2): 224–253.

Matasumoto, D. 1989. Cultural Influences on Perception of Emotion. *Journal of Cross-Cultural Psychology,* 20(1): 92–105.

McClelland, D. C., Atkinson, J. W., Clark, R. A., and Lowell, E. L. 1953. *The Achievement Motive.* Norwalk, CT: Appleton Century-Crofts.

McDermott, W. 2016 (November). SAP's CEO on Being the American Head of a German Multinational. *Harvard Business Review,* pp. 35–38.

McLoughlin, C. 2001. Inclusivity and Alignment: Principles of Pedagogy, Task and Assessment Design for Effective Cross-Cultural Online Learning. *Distance Education,* 22(1): 7–29.

Meissonier, R., Houzé, E., and Lapointe, L. 2014. Cultural Intelligence During ERP Implementation: Insights from a Thai Corporation. *International Business Research,* 7(12).

Meyer, E. 2014. The Culture Map: Breaking Through the Invisible Boundaries of Global Business. Hachette Book Group, Imprint of Public Affairs Books. ISBN-13:9781610392501. https://www.publicaffairsbooks.com/titles/erin-meyer/the-culture-map/9781610392501/

Meyerson, D., and Martin, J. 1987. Cultural Change: An Integration of Three Different Views. *Journal of Management Studies,* 24(6): 623–647.

Minkov, M. 2011. *Cultural Differences in a Globalizing World.* Bingley, UK: Emerald Group Publishing Limited.

Molinsky, A. 2013. *Global Dexterity: How to Adapt Your Behavior Across Cultures Without Losing Yourself in the Process.* Boston, MA: Harvard Business Review Press. ISBN-10: 1422187276 | ISBN-13: 978-1422187272.

Mustafa, G., and Lines, R. 2012. The Triple Role of Values in Culturally Adapted Leadership Styles. *International Journal of Cross Cultural Management,* 13(1): 23–46.

Newman, K. L., and Nollen, S. D. 1996. Culture and Congruence: The Fit Between Management Practices and National Culture. *Journal of International Business Studies,* 27(4): 753–779.

Ng, K. Y., Tan, M. L, and Ang, S. 2011. Global Culture Capital and Cosmopolitan Human Capital. In A. Burton-Jones and J. C. Spender (Eds.), *The Oxford Handbook of Human Capital,* pp. 96–119. Oxford, UK: Oxford University Press.

Nisbett, R. E. 2003. *The Geography of Thought: How Asians and Westerners Think Differently . . . And Why.* New York: Free Press. ISBN-10: 0743255356 | ISBN-13: 978-0743255356.

Osland, J. S., Bird, A., Delano, J., and Jacob, M. 2000. Beyond Sophisticated Stereotyping: Cultural Sensemaking in Context. *Academy of Management Executive (1993–2005),* 14(1): 65–77.

PMI. 2013. *Managing Change in Organizations: A Practice Guide.* Newtown Square, PA: Project Management Institute.

Pogosyan, M. 2017 (February 21). Geert Hofstede: A Conversation About Culture. Retrieved February 19, 2018, from https://www.psychologytoday.com/us/blog/between-cultures/201702/geert-hofstede-conversation-about-culture

Ralston, D. A., Egri, D. E., Stewart, S., Kaicheng, Y, and Terpstra, R. H. 1999. Doing Business in the 21st Century with the New Generation of Chinese Managers: A Study of Generational Shifts in Work Values in China. *Journal of International Business Studies,* 20(2): 415–428.

Ralston, D. A., Gustafson, D. J., Terpstra, R. H., and Holt, D. H. 1995. Pre-Post Tiananmen Square: Changing Values of Chinese Managers. *Asia Pacific Journal of Management,* 12: 1–20.

Ramsey, J., and Lorenz, M. P. 2016. Exploring the Impact of Cross-Cultural Education on Cultural Intelligence, Student Satisfaction and Commitment. *Academy of Management Learning and Education,* 15(1): 79–99.

Rogers, C. P., Graham, C. R., and Mayes, C. T. 2007. Cultural Competence and Instructional Design: Exploration Research into the Delivery of Online Instruction Cross-Culturally. *Educational Technology Research and Development,* 55(2): 197–217.

Samovar, L. A., Porter, R. E., and Jain, N. C. 1981. *Understanding Intercultural Communication,* p. 24. Belmont, CA: Wadsworth.

Schein, E. H. 1992. *Organizational Culture and Leadership*. San Francisco, CA: Jossey-Bass.

Schneider, S. C., and Barsoux, J. 2003. *Managing Across Cultures*. Harlow, UK: Financial Times Prentice Hall.

Schultz, S. 2009. *Why Culture Matters—An Empirical Study of Working Germans and Mexicans: The Relationship Between National Culture, Resistence to Change and Communication*. Herzogenrath / Maastricht, Germany: Shaker Verlag GmbH. ISBN-10: 3832278516 | ISBN-13: 978-3832278519.

Schwartz, S. H. 2008. *Cultural Value Orientations: Nature and Implications of National Differences*. Moscow: Publ. House of SU HSE.

Smith, A., and Graetz, F. M. 2011. *Philosophies of Organizational Change*, p. 32. Cheltenham, UK: Edward Elgar Publishing Inc.

Steensma, H. K., Marino, L., and Dickson, P. H. 2000. The Influence of National Culture on the Formation of Technology Alliances by Entrepreneurial Firms. *Academy of Management Journal*, 43(5): 951–973.

Tannen, D. 1983. The Pragmatics of Cross-Cultural Communication. *Applied Linguistics*, 5(3): 189–195.

Thomas, M., Mitchell, M., and Joseph, R. 2002. The Third Dimension of ADDIE: A Cultural Embrace. *TechTrends*, 46(2): 40–45.

Triandis, H. C. 1986. Collectivism vs. Individualism: A Reconceptualiztion of a Basic Concept in Cross-Cultural Social Psychology. In C. Bagley and G. K. Verma (Eds.), *Personality, Cognition and Values: Cross-Cultural Perspectives of Childhood and Adolescence*. London: Macmillan.

Triandis, H. C. 1989. The Self and Social Behaviour in Different Cultural Contexts. *Psychological Review*, 96, 269–289.

Triandis, H. C. 1994. Cross Cultural Industrial and Organizational Psychology. In H. C. Triandis, M. D. Dunnette, and L. M. Hough (Eds.), *Handbook of Industrial and Organizational Psychology*, pp. 103–172. Palo Alto, CA: Consulting Psychologists Press.

Triandis, H. C. 1995. *Collectivism and Individualism*. Boulder, CO: Westview Press.

Trompenaars, F., and Hampton-Turner, C. 2012. *Riding the Waves of Culture* (3rd Edition). New York: McGraw-Hill Education. ISBN-10: 0071773088 | ISBN-13: 978-0071773089.

Whelan-Berry, K. S., Gordon, J. R., and Hinings, C. R. 2003. Strengthening Organizational Change Processes. *The Journal of Applied Behavioral Science*, 39(2): 186–207.

Wibbeke, E. S., and McArthur, S. 2013. *Global Business Leadership*. London, UK: Routledge.

Young, P. A. 2008. The Culture Based Model: Constructing a Model of Culture. *Educational Technology and Society*, 11(2): 107–118.

Yunlu, D. G., and Clapp-Smith, R. 2014. Metacognition, Cultural Psychological Capital and Motivational Cultural Intelligence. *Cross Cultural Management,* 21(4): 386–399.

Zoogah, D. B., and Abbey, A. 2010. Cross-Cultural Experience, Strategic Motivation and Employer Hiring Preference: An Exploratory Study in an Emerging Economy. *International Journal of Cross Cultural Management,* 10(3): 321–343.

Further Reading

Adsit, D. J., London, M., Crom, S., and Jones, D. 1997. Cross-Cultural Differences in Upward Ratings in a Multinational Company. *The International Journal of Human Resource Management,* 8: 385–401.

Ang, S., and Van Dyne, L. 2008. Conceptualization of Cultural Intelligence: Definition, Distinctiveness, and Nomological Network. *Handbook of Cultural Intelligence: Theory, Measurement, and Applications,* pp. 3–15. Boca Raton, FL: Taylor & Francis.

Ang, S., Van Dyne, L., and Tan, M. 2006. Personality Correlates to the Four-Factor Model of Cultural Intelligence. *Group and Organization Management,* 31: 100–123.

Berry, J. W. 1980. Acculturation as Varieties of Adaptation. In A. M. Padilla (Ed.), *Acculturation: Theory, Models and Findings,* pp. 9–25. Boulder, CO: Westview.

Crowne, K. A. 2008. What Leads to Cultural Intelligence? *Business Horizons,* 51: 391–399.

Fortune Magazine. 2017. Fortune Global 500. Accessed from http://fortune.com/global500

Hmielski, K., and Ensley, M. 2007. A Contextual Examination of New Venture Performance: Entrepreneur Leadership Behaviour, Top Management Team Heterogeneity, and Environmental Dynamism. *Journal of Organizational Behavior,* 28(7): 865–889.

Huang, L., Lu, M. T., and Wong, B. K. 2003. The Impact of Power Distance on Email Acceptance: Evidence from the PRC. *Journal of Computer Information Systems,* 44(1): 93–101.

Javidan, M., and Bowen, D. 2013. The Global Mindset of Managers: What It Is, Why It Matters, and How to Develop It. *Organizational Dynamics,* 42(2): 145–155

Lazar, O. 2011. *Ensure PMO's Sustainability: Make It Temporary!* Paper Presented at PMI® Global Congress 2011, North America. Newtown Square, PA: Project Management Institute.

Lazar, O. 2016. *When Change Is Not a Change Anymore: Organizational Evolution and Improvement Through Stability.* Paper Presented at PMI® Global Congress 2016, EMEA, Barcelona, Spain. Newtown Square, PA: Project Management Institute.

Luthans, F. 2002. Positive Organizational Behaviour: Developing and Managing Psychological Strengths. *Academy of Management Executive,* 16(1): 57–72.

Luthans, F., and Youssef, C. M. 2007. Emerging Positive Organizational Behaviour. *Journal of Management,* 33(3): 321–349.

Luthans, F., Avolio, B. J., and Youssef, C. M. 2007. *Psychological Capital.* New York, NY: Oxford University Press.

Luthans, F., Avolio, B. J., Avey, J., and Norman, S. M. 2007. Psychological Capital: Measurement and Relationship with Performance and Satisfaction. *Personnel Psychology,* 60: 541–572.

Nisbett, R. E., and Masuda, T. 2003 (September 16). Culture and Point of View. *Proceedings of the National Academy of Sciences of the United States of America.* https://doi.org/10.1073/pnas.1934527100

Offermann, L. R., and Hellmann, P. S. 1997. Culture's Consequences for Leadership Behavior. *Journal of Cross-Cultural Psychology,* 28(3): 342–351.

Triandis, H. C., Bontempo, R., Villareal, M., Asai, M., and Lucca, N. 1988. Individualism and Collectivism: Cross-Cultural Perspective on Self-Ingroup Relationships. *Journal of Personality and Social Psychology,* 54: 323–338.

Index